The World of the Tent-Makers

Drawings by Abigail Rorer

The University of Massachusetts Press

Amherst, 1980

THE WORLD
OF THE TENT-MAKERS

A Natural History of the Eastern Tent Caterpillar

Vincent G. Dethier

Library of Congress Cataloging in Publication Data
Dethier, Vincent Gaston, 1915–
The world of the tent-makers.
Bibliography: p.
1. Eastern tent caterpillar. 1. Title.
QL561.L3D47 595.78′1 80–11361
ISBN 0–87023–300–9
ISBN 0–87023–301–7 pbk.

Contents

Thrice happy he who, not mistook,
Hath read in Nature's mystic book.
Andrew Marvell—"Upon Appleton House"

I

A chill no coat, however stout,
of homespun stuff could quite shut out, . . .
Whittier—"Snowbound"

An old-fashioned New England blizzard raged into its
second day. Nearly two feet of snow had fallen before the storm
blew itself out on the morning of the third day. With its passing a
keening wind whipped the snow into shoulder-height drifts,
rounded and cornised. The last scud disappeared over the eastern
horizon. From a flawless blue sky the sun sought out billions of
facets in the snow crystals. Over this sparkling world dropped a
cold that drove the thermometer to twenty degrees below zero.

Some life stirred. The farmer emerged from his house in the valley and began to shovel his way to the barn. His red-jacketed figure
bobbed rhythmically with each shovel-full which the wind plumed
away at the end of the swing.

In the white pines hardy chickadees and nuthatches that were
not patronizing the backyard feeder foraged optimistically. King-
lets visiting from the north busied themselves in the hemlocks.
Bluejays and hairy woodpeckers contested for suet hung in a bag
near the house. Gray squirrels dug down through the fresh snow in
search of cached acorns. Most other life lay dormant in some well-
chosen protected place. Far beneath the sod in the apple orchard the
woodchuck slept a dreamless sleep, oblivious of the frigid world
above. The black bear kept to his den in the hills. Frogs and toads
squatted fossil-like in the mud of the swamps. Perhaps in some
future geologic age those that did not survive would indeed be
fossils. Myriads of insects lay mummified beneath the meadow
sod, in and under logs, in crevices in the bark, even in unheated

barns and attics. All in their own way had sought shelter well in advance of winter.

Yet there were some that neither slept in the protection of the ground nor fluffed their fur or feathers in defiance of the cold. Some few remained exposed to the weather in all its moods. In an abandoned orchard ancient apple trees creaked and groaned like the arthritic old countrymen who could no longer tend them. Beneath the hoary bark of the trunks and the smooth thin bark of twigs alike the live ring of cells of the cambium, their growth temporarily arrested, nevertheless lived. Cold as the sea of polar air around them, they lived. Stands of choke cherries, many disfigured by black knot, thrust up through the drifts along the roadsides, at the edges of fields, and bordering coppices. Their cells too, exposed and unprotected, remained alive in the bitter cold as did the virus of the black knot that parasitized them.

Still other creatures, equally exposed, remained alive. Here and there on the outermost twigs of the apples and cherries a sachet of minute eggs challenged the cold. Packed side-by-side, their only blanket was a poor thin layer of shellac-like froth, an effective protection against the desiccating winds of a summer but no insulation against the frigid air of winter. Each egg, the likeness of a tiny porcelain thimble, contained a miniature caterpillar, immobile, yet indisputably alive.

In what would now appear to have been arrant folly the eggs had been glued in place by an undistinguished-looking moth scarcely three days old at the time. She that had never experienced cold left her offspring a legacy of winter. From a cocoon fastened to the edge of a shingle on a weathered barn she had struggled free into the warmth of the slanting rays of a four o'clock June sun. Others of her kin had emerged from similar yellowish cocoons hidden in cracks and crevices of tree trunks, field-stone walls, corn cribs, and houses. Creatures of warmth and darkness, myriads of moths took to their wings. Summer was still young. By half past seven the sun had set. The tent caterpillar moths had commenced their brief lives. They had no memory of winter. They had no prescience of winter.

Crammed into that first night of life were the experiences which

other animals live in weeks or months or years. Time was short; a destiny was to be fulfilled. As soon as her wings were unfolded, stretched and tempered with drying, the moth was mature. She experienced no adolescence, no learning. She flew without trial, without instruction, without physical conditioning. She flew a sustained flight of thirty minutes without pausing to rest. When she did rest, the respite lasted scarcely fifteen minutes. Another flight, a rest, a flight, and so on till morning. By the time she finally stopped, the farmer had milked the cows, completed his early morning chores, and had gone back into the house for breakfast. The birds had finished their morning chorales, and the bees were hard at work in the meadows. It was nine o'clock.

During her brief rests the moth had arched her abdomen to the night breeze and dispensed an alluring scent, a scent of which other creatures were unaware, a scent that held meaning only for males of her own species. Sometime during that night her call was answered. She was courted and mated.

The day was a time for rest. While she clung to a branch of apple, camouflaged in the sight of all but the keenest eye, the daytime world bustled about its business. Some of her kind were discovered by predatory beetles; some fell prey to sharp-eyed jays. She survived the dangers of day, and by evening once again took to wing. This night was not for courting. The advances of eager males were spurned. Once courted, once mated, her senses and brain were now tuned to other stimuli. She scouted the hedgerows; she entered the orchards. On the cherries and apples, guided by subtle botanical scents, she laid her eggs. The next day, unequipped by nature to eat, her life fulfilled, she died. Perhaps she was picked off in her feebleness by a quick-eyed vireo. Perhaps, falling to the ground, she was spied by a foraging woodmouse, or her corpse was dragged away and dismembered in the name of communal living by scavenging carpenter ants.

What a strange manner in which to leave a legacy, never nurtured in the womb, abandoned without concealment, with scant protective cover, a froth that by winter's end would be frayed and eroded. A poor garment indeed—homespun stuff that could not quite keep out the cold.

These were the eggs that were tossed and buffeted by January's blizzards. Against all common sense, as if to compound the folly, the eggs had not waited in a state of undeveloped dormancy, simple cells insulated and resistant in thickened walls. They had not postponed their development over the cold season as had the legions of summer seeds entombed in the frozen ground or the stubbornly resistant spores of protozoans and fungi. Spores, seeds, cysts in uncounted numbers waited for the urging warmth of spring—a most sensible time to embark upon life.

Or, there were those animals that, racing the sun in its flight to the winter solstice, developed as far toward adulthood as they were able. They garnered every calorie they could even into Indian Summer, and then sought shelter. The cutworms in the garden burrowed into the soil where they molted into sleek brown pupae. Woolly bears hurried across the highways and byways and curled up among the leaves which autumn was fast providing. By November every living thing, it seemed, had prepared for winter in its own way. The last robin had long since departed. The little brown bats had forsaken the barn and migrated to distant caves. The farmer had filled his silo. The hay was in, the squash gathered, the kindling split.

Over the summer, the eggs that the tent caterpillar moth had laid so carefully and abandoned so cavalierly in the most exposed of sites had been to all outward appearances infertile. With the long, warm months ahead of them, they had done nothing. More than one hundred years ago the entomologist Le Baron remarked on the mystery. "How is it," he wrote, "that these little germs of being remain insensible to the heat of July, August and September, and yet burst into vitality on almost the first touch of spring? We know that if the young caterpillars came out in the fall, they would perish from inability to eat the tough autumnal foliage. But what natural law can we conceive of that exercises such discretionary power?"

July passed, and August, and then September. All about them other early eggs had hatched, but these seemed to miss the golden opportunity provided by summer warmth and gloriously long days. So it was that winter found them.

When you think of eggs, you may envision fragile creations, pre-

cious promises of life, tiny bits of creation demanding the utmost care, and indeed worthy of it. You think of warmth, of tepid waters, of incubation. There are, however, the eggs of autumn, eggs laid at the end of summer when time has run out on life. Seek in the soil, and there you will find the eggs of next summer's crickets, minute promises of song in an earthern crypt. If you skilfully dissect any of these eggs, you will find—an egg. You will discover a few cells and yolk, no sentient individual, just a commitment. Dissect the egg of the tent caterpillar moth at this season and you find a caterpillar. Far from having been infertile the egg is really no longer an egg. It is a shell containing a caterpillar, fully formed, curled in a half twist. It has been there all summer. It is there when winter arrives.

Outside the wind that followed the blizzard lashes the cherry branch. The temperature still hovers at twenty degrees below zero. Inside the egg shell the temperature of the caterpillar is also twenty degrees below zero, but it is not frozen solid!

Somehow it has defied the cold. It seems to have violated a fundamental law because it, like all creatures, is mostly water. The brook in the copse is icebound and silent. In the garden the brussel sprouts that stick up through the snow are frozen solid. Where sap has dripped from a wound in the maple, a long brown icicle hangs. When the farmer drives his pung home from the woodlot, his beard is stiff with frozen breath. The horse's breath, too, has frozen on the hairs of her mouth. Only those warmblooded animals whose biochemical fires keep them fluid and supple move in the cold with their accustomed grace and agility. Even those that hibernate have only banked their fires so that in the chill of their dens they sleep unfrozen. Wherever there was water without heat there is now ice—almost without exception. The tent caterpillar is one exception. In the delicacy and vulnerability of its youth, fully formed but as yet unhatched, it remains soft and unfrozen.

The sword of winter is the ice crystal, exquisitely forged but lethal. Living cells are mostly water. When that water freezes, the growing ice crystals, fabricated around nuclei of dirt, flaws in smooth surfaces, edges or particles of molecular dimensions, stab, tear, and lacerate the fine biological machinery. What they do not destroy by force, they kill more subtly by removing water and leav-

ing behind concentrated solutions of salt with which the finely tuned molecular machinery of the cell cannot cope. Freezing water in spaces between cells can even draw so much water from the interior that the cells literally dry up to the point of being irreparably damaged. Even were they to survive freezing, the trauma of thawing rivals the destruction of freezing.

Distortion of the harmony of delicate structure, intolerable concentrations of the body's chemicals, and desiccation—these are the dangers; and the culprit is water, that most necessary of compounds, that precious commodity that sets Earth apart from its sister planets.

Some small creatures are able to defy the crystalline swords of freezing water. The crystals stitch their way through the small bodies, drawing closer and closer together, like a darning needle closing the hole of a sock with overlapping and crisscrossing threads. Finally no water remains, no space remains, the creatures are as solid as the icebound brook. Like the brook in spring they thaw, none the worse for their experience.

Other creatures cannot tolerate the crystals. They should all die, but many do not. Nor do they freeze. It is as though, for them, a magic spell is cast. And in a manner of speaking it is. The magic has only a quite ordinary name. It is supercooling.

If the water is cooled very slowly, with extreme care, and with great precaution to prevent the formation of any nuclei upon which ice crystals can form, the water can be dropped to a temperature far below that which would ordinarily cause freezing. Down, ever down the temperature drops, minus five, minus ten, minus twenty, and the water still flows. Touch it with ice, however, and the spell is broken. Instantly it freezes solid.

The unhatched tent caterpillars have made use of the trick of supercooling. They have adjusted also to the fact that the magic requires long preparation. Even before the goldenrod, harbinger of fall, has burnished the fields, before the second crop of hay is harvested, before the goldfinches have begun to nest, and before they or any other birds have given a thought to migration, the caterpillars have begun their preparation. Well before autumn arrives, subtle but mysterious changes in their blood and protoplasm have

given them the potential to survive. The cooler days of October and the gradual advent of crisp November nights hasten the process. Cold protects against cold so that the caterpillars are cold-hardy before the first freeze. By the time the frost is on the pumpkin they have the ability to withstand temperatures as low as forty degrees below zero. Intolerant, not of cold, but of freezing, unable to survive its tearing and drying, they can, nonetheless, withstand the most numbing cold that winter can bring.

We know little of what the biochemical machinery does to prevent the freezing. Partly it rids itself of some excess water; partly it manufactures an antifreeze. Millions of years before man discovered antifreeze, or indeed even existed, tent caterpillars and some other animals of the northlands were making it for their own protection. Glycerol not only depresses the freezing point of body fluids, it greatly lowers the supercooling point. In summer one percent of the unhatched caterpillar is glycerol. By Thanksgiving the ratio has risen to nearly twenty percent, by January it tops thirty-five percent, where it remains until April. With the returning warmth of spring it falls rapidly. A continuing warm spell of nine days drops it below its summer level. What need of an antifreeze now, or ever more? Some mystery remains. In summer the supercooling point begins to fall before glycerol makes its appearance in the blood. It cannot be the only antifreeze.

Just beneath the floor of the egg the cherry twig has wrought some of the same magic to stay alive. It too, like other northern trees, began to build a cold-hardiness before the approach of winter. Its cells change their properties as mysteriously as do the caterpillar's cells and they also lower their supercooling point. Water is removed; the cell sap is concentrated. Between cells the fluid might freeze, but no crystals penetrate the cell walls.

The sun dropped low on the first bitter day of winter. Wan rays fingered through arched birches. Gradually the wind hushed to a whisper. In the corn field a tractor, left out till spring, hunched in its own snow drift. Its engine had been drained. Like the seeds and spores in the soil beneath it, it had no water. It would not freeze. The truck in the barn would not freeze either. Its radiator was filled with antifreeze—not glycerol, but a chemical relative, glycol.

By half past four the wind was breathing its last. The air held no fragrance except the smell of wood smoke drifting up from the farm in the valley. The temperature which had climbed to zero during the day settled back down to twenty below. A maple limb, weakened in the storm, and frozen, cracked like a pistol shot. Misshapen oak leaves, rattling brown skeletons, their vital fluids dried weeks before, clung stubbornly to the trees. All that characterized life, fluid motion, softness, suppleness, and warmth was absent. To the unknowing, nothing alive appeared to exist in the broad expanse of white. There were no sounds. Squirrels had retired to the protection of their leafy shelters. The chickadees huddled in the pines. The caterpillars grew colder with the night. Were they really alive? What in the final analysis was life? Being or potential?

Overhead the constellation Orion shone clear and bright. Many cold nights and days lay ahead. Spring was still a long way off.

II

The third day comes a frost,
a killing frost.
Shakespeare—*King Henry VIII*

Toward the end of January the mid-winter thaw came. Weather bureau statistics said there was no such thing. The fact remains, however, that during the night the thermometer shot up to the high sixties. The black blowflies wintering under the shingles of the barn stiffly crawled out to bask in the sun. The coarsening snow became peppered with springtails, those weird microscopic creatures that get into the maple syrup buckets in spring. At night some pale geometrid moths came out of hiding and danced about the porch light as though full summer had burst upon them. Those furry hibernators who slept less soundly than others emerged for a stroll in the mushy snow. The tracks of a skunk ambled up to the edge of the road only to disappear in the water rushing down the gutters and middle alike. The footprints did not reappear on the far side.

Water stood or flowed everywhere. Roofs shed their burdens of snow. Gutters unfroze. Drain spouts echoed metallically to the tattoo of falling drops. An unaccustomed sound was borne from the woods on a warm breeze—a roar, where before only silence had been. The rising temperature had unbound the brook. All the ponds became ponds again, shallow ponds with thick ice bottoms. Everywhere the snow was a worn and soiled blanket. In sheltered spots rosettes of dandelions flaunted an incongruous green in all the desolation of grays and browns. Other dormant herbs, sorrel, Queen Anne's lace, and great plantain, still hugged the ground but also added their bit of green. The moist southern air seemed to in-

tensify the yellow of the bare-branched willows and the crimson of the red osier along the brooks. Many living things did not respond. The false spring interlude had no more effect upon the tent caterpillar eggs than had the warm days of late summer. The sun at the highest point of its modest winter arc drove the internal temperature of the eggs up to seventy-five degrees. Inside the shells the microclimate approached that of a warm midsummer day. One might have imagined that the young caterpillars would become restless, would squirm in their prison cells, would, in a fit of impatience, gnaw their passage to freedom. What a catastrophe if they would have emerged! No food could have been found. Even if it had, winter was playing tricks. Only the unwary were deceived. The blizzards of February still lay ahead, also the spring storms of March. Unlike the geometrid moths carousing around the light and able to go back into winter sleep, the caterpillars once out could never crawl back into the egg. There is no return to the womb and no rebirth. Some wisdom of nature, evolved over the millennia, protects them. They did not awaken to the rising temperature any more than they had stirred in the warm days of Indian Summer.

Perhaps the respite from winter is too brief. Cut a cherry twig with its burden of eggs. Bring it into the warmth of the house where it can stand safely watered in a tumbler or vase. The days pass, and the weeks. There is no sign of life from the small brown packet. Nor does the twig itself burst into bud. Finally, after nearly two months, a few caterpillars emerge into daylight, explore the twig for leaves that are not there, then huddle together and slowly starve. A few weeks later the cherry puts forth leaves. Premature warming has forced nature's clocks out of synchrony. The earlier in the year, the more marked the dislocation. If the eggs are brought indoors at Thanksgiving, one can wait in vain for signs of life.

At the time of the January thaw heat will seduce neither the caterpillars nor the cherry. Paradoxical as it may appear, the very cold that arrests hatching in the fall to protect the caterpillars against winter is an absolute requirement for hatching when winter departs. Unless the eggs experience from three to five months of cold, just above freezing, few of them will hatch when exposed to warmth. At lower or higher temperatures a still longer period of

cold-conditioning is required; and the more incomplete the conditioning, the longer the eventual delay in hatching. What a beautiful design for survival: a deep winter's sleep from which no siren zephyrs from the south can awaken them prematurely.

The unhatched tent caterpillars are not the only ones that sleep the enchanted sleep, the sleep of diapause. In the insect world the obligatory rest may come at any stage of life, to the egg, the larva, the chrysalis, or even the adult. Unlike the lethargies and dormancies of many creatures, the geometrid moths, the flies, and those hibernators like the skunk, unlike the light sleep from which the warmth arouses, the deep sleep is beyond disturbing.

It is a curious sleep, this diapause. It begins before the cold sets in, it requires weeks of winter before it can be disturbed, and warmth will not terminate it before the appointed time. It is as though at some signal nature sets an alarm clock in midsummer, and marks off the time from that setting. What can the signal be, and how is the time reckoned? If the signal is not temperature, what can it possibly be? Surely temperature is a false and variable friend. It would awaken the caterpillars in the mid-January thaw. It would lure them out when March comes in like a lamb only to smite them when March goes out like a lion. No, the alarm must be less fickle. It must be something with a sure margin of safety, something with the constancy and precision of time. It must be time itself.

Someday perhaps, our children or our children's children will be carrying atomic wrist watches as people today carry digital watches that display the time at the touch of a button. Someday, perhaps, they will have forgotten the watch in which the hands sweep inexorably around the dial, displaying the continuity and the passage of time. Today the hourglass is only a toy, and the sundial is all but forgotten. Long before their invention, men marked the time of the seasons with the sun stone, the "Friar's Heel" at Stonehenge, and even before that they read the changing course of the sun and the circling of the constellations. The sky was their calendar. There, time and the cycle of the seasons is marked, not by the fickle vagaries of temperature, but by the precise procession of the heavenly bodies. In the heavens is to be found nature's clock.

Few of the earth's lesser creatures can scan the passage of the stars. Many have neither the vision to see the stars nor the brains to fathom their meaning. Many are shut away from a view of the sky. Yet, eyes or no, impoverished is the creature that is insensitive to light and dark. Light penetrates the egg and the cocoon. The skin itself is no barrier. Somewhere in the caterpillars' bodies the cycle of day and night is sensed. Somewhere within them are their own biological clocks marking the passage of time.

The clock is set just after the sun has reached the summer solstice. In southern New England the earliest eggs are laid during the nights of early June. Since the caterpillars require three or four weeks to develop, their little brains are too formless to have any midsummer night's dreams even were they able. The longest day passes unheeded. Even though the bounty and fertility of summer still lies ahead, and all of nature is exploding in paroxysms of growth, the sign of winter is already in the sky. Nature's summer may be only beginning, but the days have already begun to shorten.

Through the translucency of the egg shells the caterpillars receive each day just a little bit less light. Their period of incubation is finished. By rights they should hatch; however, the shortening days and lengthening nights start to set the mysterious biological clock. The passage of time and of the seasons is marked independently of temperature. The clock shuts off the machinery of development. Four weeks of progressively cooler autumnal days and nights complete the cold-hardening preparation for winter. The caterpillars are conditioned, and the clock has been set for three months of cold storage. Nothing can now awaken them until the clock has marked off the appointed time.

Do not be deceived, however, by our kinds of alarm clocks which, set in the evening, jangle us out of sound slumber in the morning. There is no insistent jangling for the caterpillars. The clock is permissive rather than commanding. The shortening days set it to measure off the winter months, and it measures most quickly when temperatures hover around freezing. If there is no cold, the alarm never goes off. When it does go off, it does not arouse the caterpillar. It simply tells the caterpillar that its sleep is ended. It permits the caterpillar to awake. If the weather continues

cold, the laggard lies abed without stirring. One need not arise just because the alarm rings.

One might expect that the lengthening days would break the long winter sleep just as the gloaming of fall induced it, but a "light" setting, though a friend in autumn, can be a traitor in spring. In northern latitudes changes of seasonal temperature lie behind changes in length of day. Were the caterpillars to emerge in slavish response to light, they could very well be killed by late March freezes. A week of basking in warm sunlight is needed to stir them to activity. This second incubation allows a few finishing touches to be made in development. It perfects some roughness that does not meet the eye. It drives off the last vestiges of cold.

Before this time it may seem a kindness to bring the caterpillar in from the cold. One can even help it carefully out of the egg, but it refuses to awaken. Like any deep sleeper it responds fretfully to insistent prods. As far as can be seen it is completely formed; it twitches, and presumably itches; but if uncurled, it slowly returns to its cradled posture like a reluctant spring. One cannot help but wonder what is going on in the small brain.

Although caterpillars do not have the same kinds of brain waves that we have, their waking brains are maelstroms of electrical activity. Thousands of cells are discharging tiny voltages, the language of the nervous system. They are constantly receiving messages from the outside world and from the goings on of their own bodies; they are making decisions and issuing commands. It is tempting to believe that the caterpillar rudely removed from its egg does not "come to life" because its brain is inactive. The truth is that the brain is ticking steadily, and the subtle biochemical systems that keep it going are as fully developed now as later.

Yet, an alertness is missing. A rude poke in the midriff is felt. A message travels from the sense organs of touch to the brain, which in turn sends a half-hearted message back down to the muscle, saying "twitch." There is no message saying "crawl away from that annoying thing that is poking." In the egg, dim light and shadows filter weakly through the shell. Released, however unwillingly, the caterpillar lies now in a flood of light. The six small eyes on each side of the head become brightly illuminated. Their connections to

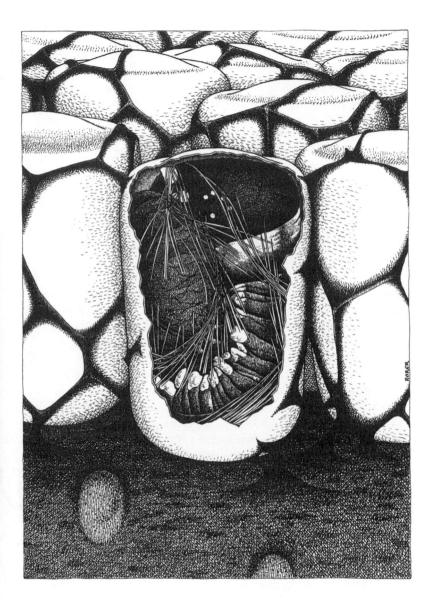

the central nervous system are fully formed, but there is no command to crawl either toward or away from the light. The brain is far from being a sleeping brain, but it is certainly inattentive and lackadaisical.

There is one system, however, that does appear to be truly idle. This is the digestive tract. At first thought inactivity would appear to be quite logical. After all, the caterpillar has yet to eat its first meal, but it has inherited a meal from its mother. While it was being formed within the egg, a large mass of yolk became enclosed by the embryonic digestive system. As a consequence, in early winter the gut is tightly packed with yolk, the only food available till hatching. As cold storage stretches into the third month of low temperatures, the yolk remains undigested. Only when the approach of spring is heralded by ever-increasing intermittent periods of warmth will the cells of the digestive tract become active, begin to digest the yolk, and supply energy not only to maintain the small body in its dormant state, but also to prepare it for the exertions of hatching. At the same time the excretory system, of little use during the winter, will begin to prepare itself for the active life to come.

All of these transformations lie in the future. The midwinter thaw is powerless to initiate them. A prolonged basking in the comfort of a house is equally ineffective. The clock has not run through its set time. Many more weeks must pass, but the day will come when the sun shining on the egg will awaken the caterpillar. A week of incubation will complete the preparation for a strenuous independent life. We do not know what subtle changes allow these matters to proceed after several months of cold treatment but not before. We do not know where the clock is or what it is. Nor do we understand how light sets it and cold regulates it. We are aware only that the hormonal tides that ebb and flow during different stages of life rise to a flood at a time that slightly anticipates spring and unlock the restraints on development.

Length of day, frosts of winter, warmth of spring—these are the signals that turn on and shut off the hormones in an intricate interplay between the outside world and the inside. Events are so finely tuned, evenly balanced, checked and counterchecked that vagaries

of weather cannot trick the caterpillars into premature and ill-advised actions. Nature protects against its own caprices.

The mid-January thaw will not delude the caterpillars. It will not entice them from the eggs and abandon them, orphans, to the blizzards yet to come. The polar air from Canada will return. The geometrid moths will crawl under the bark and shingles, the blowflies will return to the nooks and crannies, the skunk will turn his back on the moon and return to his home. By the third day a killing frost will return. Again the brook will become silent.

III

The stress that winter can impose on the eggs is indeed severe. Not many days after the thaw had ended and the temperature had plummeted to a more seasonable low a vast river of chill moisture flowed down the valley and piled up and over the hills. These winter fogs strike to the very marrow.

Past midnight and during the early hours before sunrise the undercooled particles of fog collided against all exposed objects. The rural mailbox, the poles supporting the clothesline and the clothesline itself, the weathercock on the barn, the topmost and edgemost branches of trees and shrubs, all these exposed things that had radiated away their heat became freezing cold. The breeze piled the fog particles upon them, where each microscopic drop crystallized. Gradually on the windward sides of things jagged needles and fingers and pinnacles of etched ice began to build in an unbelievable horizontal gothic. Millimeter by millimeter, centimeter by centimeter, this rime thrust out into the wind, a streamlining in reverse, white pennants pointing into the wind instead of away from it. By the time the sun had arisen sparkling on the rime everything that projected above the relict snow had been exquisitely mantled. The dried umbles of the Queen Anne's lace of yesterseason, the spindly stalks of St.-John's-wort, the spent and coarsened skeletons of the goldenrod's blossoms, the wizened shards of empty milkweed pods, these and all others recaptured an ephemeral splendor. The cherry too, and the tent caterpillar eggs as well, had squandered the heat that they had acquired during the thaw; and the rime, not dis-

criminating the living from the dead, built its icy castles on the cold eggs.

On many more occasions before the winter was to end, the eggs, the cherry, and all other surfaces of nature would become colder than the air around them. One morning when this happened a driving rain began to fall. As fast as the drops flattened on the ground they froze. Within hours the ice storm had glazed everything as far as the eye could see. All of nature's wrinkles and rough spots were given the smoothness of youth. All of man's imperfections, the roughness of shingles, the angularities of buildings, the worn treads on doorsteps, the coarseness of walks, were smoothed and polished to a fault. No longer did edges exist. There were no discontinuities to be found anywhere. The world was one, everything bonded in an all-embracing mantle of ice which grew even thicker and heavier.

Under the crystal burden the lesser trees and bushes bowed to the ground. Shrubs and grasses, slight supple things, capitalized on their resilience—and survived. The gray birches tried. When they had touched their crowns to the ground in obeisance to the storm, they could humble themselves no more and broke under their burden. The old apple trees in the orchard had long since lost the litheness of youth. Many were shattered—one of nature's ways of removing the aged to make way for the young. The eggs on the outermost branches would survive their glazing of ice; but when spring arrived, the young caterpillars would find no nourishment on the fractured dead branches, and would die. Those on the younger cherries would fare better. They too were completely encapsulated in a crystal sarcophagus, but the cherry did not break. Before the sun completed its transit that day the ice would have melted. Caterpillars and cherry alike would survive.

On Washington's birthday a change became noticeable. It had been developing slowly, imperceptibly, but a prod to the consciousness was required to bring full realization that winter had begun to retreat. It might have been like this in glacial times, on a vastly different time scale, of course; nonetheless, a retreat so slow that year-to-year, century-to-century changes would have passed unmarked. Just so are parents insensitive to the day by day growth of a child until suddenly a shirt is too tight or the sleeves too short

and realization dawns that growth has been going on unnoticed before their eyes. So it falls with the seasons. One awakens on a February morning. The thermometer still hovers around zero. The countryside may still lie buried under a burden of snow, but on this particular morning one suddenly realizes that breakfast is being eaten in the brilliant light of a sun which, however close to the horizon still, beggars the feeble efforts of the lamps indoors. The days have extended. There is no longer any doubt that, however many storms are yet to bear out of the northeast, winter is in retreat.

There are many signs. For one, the birds are more vocal. The same bluejays that cried "Thief! Thief!" in the fall and winter now give voice to a familiar melodious chortle. The starlings around the barn, truly talented songsters despite their ill reputations as pests and avian ruffians, have commenced warbling their imitations and variations from the tops of the barn and surrounding maples. Glancing away from the starlings for a moment and letting your eye travel from the maple crowns down the massive ribbed trunks you observe that the sap buckets have been hung. What surer sign that the new season is in the offing? The crisp morning air also carries that tentative song of the cardinal. Even more assertive is the drumming of matins by the downy and hairy woodpeckers. In the swales the willows seem more yellow and the red osiers even redder than before. Pussywillows have not yet appeared, but it is only a matter of days. All the signs are there to be heard and seen.

There are other more subtle indications. Bring a cherry twig with its cluster of eggs indoors now. In the short space of a few days, twelve at most, a lively company of caterpillars bursts forth from the egg mass. Whereas earlier in the winter you had to wait for weeks for signs of life and even then were rewarded by the appearance of only a few caterpillars, now nearly one hundred percent of the eggs hatch. The cherry too is more ready than ever. It hastens its leafing in time to be consumed by hungry hordes. The cherry's clock and the caterpillars' are now in synchrony. Only one element is wanting, a prolonged reliable period of warmth.

In the late afternoon, from their opposite horizons, the moon and the sun stare at each other across a white landscape. Change is in the air.

IV

When the hounds of spring
are on the winter's traces—
And frosts are slain and flowers begotten,
And in green underwood and cover
Blossom by blossom the spring begins.
Swinburne—"Atalanta in Calydon"

By the ides of March any lingering doubts about the demise of winter vanished with the appearance of the first migrant robins. One day there were none; the next day lively companies foraged in the fields. They appeared lean, as well they might considering how much strenuous nocturnal flying they had been putting in over the weeks past. Also, the brick-red breasts caught the sun a little less brightly than they would later when the birds became fuller. There was, however, a contradiction, rather lugubrious, in the sight of these lean birds running across the thin crusty snow and cocking their heads at intervals as though they were hunting worms on a summer lawn.

Another, even surer, sign of spring struck the ear. Winter is a silent season. No song of any sort fills the air. There are no leaves for the wind to rustle and no grass to sigh. Noises there are, but no song. By contrast, late spring and summer have their birds; midsummer and fall have their crickets, long-horned grasshoppers, and katydids. Spring has its frogs and toads. The undisputed harbingers of spring now shrilled in chorus from the meadow marshes and wooded swamps. The knee-deeps—the spring peepers or hylas—like so many of the living things "knew" that spring had arrived and proclaimed this fact to the world.

Within two weeks of the amphibian proclamation there was a great stirring among restless creatures and long-dormant plants.

Preparation for this season had, of course, been going on unperceived for a long time, but now the more adventuresome and hardy could wait no longer. Among these were the tent caterpillar eggs.

Just what moves a caterpillar to hatch at a particular moment when it has been ready but inactive for so long remains one of the mysteries. True, it had to sort out its biochemistry under the complex influences of cold, light, hormones, and warmth; but what said "start hatching this moment" or "wait two more minutes" or "put it off until tomorrow"? What, in a word, prompts the caterpillar to "make up its mind"? Is there suddenly some stimulus that a moment before was absent? Has some chemical of the body overflowed a limit and tipped a whole scale of events? Epitomized in this small beast is one of the profound mysteries of behavior in general, the mystery of what starts things.

At first only the more adventuresome emerged into the unaccustomed sunlight. The emergence combined an ordeal with a feast. By scraping away some of the froth on the egg mass one could see the first sign of activity as a very small dark keyhole in the top of the egg. At intervals this grew larger as though a door were being opened by progressively enlarging the keyhole. Soon two piceous mandibles could be recognized by their serrated edges and miniscule movements. Unlike our jaws which bite up and down, these chew from side to side.

The caterpillar labored for a few tens of seconds, then rested. Gradually the hole widened until an entire shiny head could be discerned. With additional room to maneuver the caterpillar chewed in a more businesslike manner. The accelerated pace of chewing demolished the cover of the egg in five to ten minutes. Hatching was a feast as well as an ordeal because the shell that was chewed was consumed. Life's first meal became a meal of egg shell, hardly the epitome of palatability, but surely a small source of energy to fill the gut long since emptied of yolk. Within an hour the confinement of months ended. The world invited. How hostile a world the caterpillar was soon to learn.

Where the hardened froth, the spumaline, had not been scraped from the eggs or eroded by winter storms, freedom was not achieved merely by escape from the egg shell. A double layer of

spumaline still intervened. It had not served well in insulating the eggs from the cold, but it did protect them from desiccation on dry days, summer and winter alike. The eggs themselves could become filled with water if wet, and if water could get into them it could escape. In short, eggs could dry out if not protected. Part of the protection came from the living twig beneath, but much was afforded by the spumaline which readily absorbed moisture on humid days, thus delaying evaporative loss by the eggs on dry days. For every advantage, however, there was a price to pay, the price in this case being the added difficulty of escaping the spumaline prison.

Although to the casual eye it appears as a rather uniformly grainy, shiny mass, spumaline is revealed under the microscope to be an intricately structured material resembling a mass of small leaded glass chambers. Most are hexagonal, but there is a fair representation of cubes and pentagons. The "glass" is a transparent smoky pane. The elaborate construction is less a matter of design than of physics. Anyone who has ever blown a mass of adhering soap bubbles can readily appreciate how the spumaline achieves its architectural form. It too begins as a froth of bubbles. As with the soap the bubbles arrange themselves in all shapes and sizes as they adjust almost instantaneously to surface tensions and other physical forces. Thus, the joined surfaces shared by bubbles tend to be flat and the free surfaces convex. The pushing and pulling of the forming bubbles determine the number and area of the walls formed and hence whether the enclosed space is cuboidal, pentagonal, or hexagonal. Soap bubbles soon burst into nothingness, but spumaline, being somewhat viscous, retains its form and hardens, the free convex surfaces becoming flattened and the shared walls flat and thickened. Two layers cover the eggs, providing, in theory, a double insulation of dead air space, but in fact an inconsequential barrier to cold. They are not aligned with each other, nor is the bottom one aligned with the eggs. Greater structural strength is provided by the misalignment. Through this light but rigid windowed lattice the caterpillars industriously ate a passage.

During the first two weeks of April more and more hatching occurred. Some egg masses had been placed, quite by chance, on cherries growing in sheltered locations or on southern exposures. These had enjoyed the advantage of more accumulated light and warmth

so they hatched early while others in less favored situations tarried longer. These early days of spring provided the uninterrupted span of warmth required to stir the larvae. It was not exactly uninterrupted, as later events would prove, but it lasted longer than had the mid-January thaw. The clock had long since marked the end of cold-conditioning. The temperature now hovering in the seventies completed the spring incubation. It was the drum that set the pace of marching. It brought out the caterpillars; it stimulated the budding of trees; it stirred the hylas to greater enthusiasm. It awoke others who joined in the spring chorus. Among them the common toad surpassed all in his vocalizations. Who could have guessed that such an ugly fellow could trill such pure and sustained cadenzas. He could put a cricket to shame. Not long afterwards the toads and hylas were joined by the common meadow frogs, *Rana pipiens*, which provided a low-pitched continuo.

The world had indeed come to life. Skunk cabbages thrust their tough purple augers through the swamp mud. Pussywillows replaced their kitten fuzz with the yellow pollen grains of maturity. Honeybees stretched their wings in the first flight of the new year. The songbirds in the day rivaled the frogs in the night. And the caterpillars kept hatching in waves.

Some eggs were infertile. Sperm had failed to find the small opening, the micropyle, that afforded entry to the cell inside the shell. Or the sperm itself had been impotent. Some eggs had faltered in the course of development. At some critical moment a cell had failed to divide. Or cells that should have stopped dividing when their shift was completed had gone merrily on with unseemly lack of restraint. Developmental and genetic defects ordained that some embryos would never become caterpillars. Of the one hundred to five hundred eggs per mass, from one to three percent never hatched. Even this early in life the death toll had begun. Actually the fate of many had been decided the previous summer when the embryos would normally have completed their development.

As if these were not hazards enough, death visited in another guise, in the form of highly skilled wasps no larger than an egg itself. At least five species of these Lilliputian creatures sought out freshly laid eggs, guided by an olfactory sense acute enough to rival a bloodhound. Neither the spumaline nor the tough chorion of the

egg is impervious to the microscopic ovipositors. An egg was laid within an egg. The egg of the tent caterpillar moth, minute as it is, provides all the nourishment that is necessary for the complete development of one parasitic wasp. Moreover, the wasp embryo is so tough that it can survive adversities of weather to which even the beautifully adapted host eggs succumb. Thus it comes about that spring may call forth from the packet of eggs not caterpillars but wasps. Only two to nine percent of the eggs may be parasitized, but together with the developmental defects and adverse weather the toll begins to mount.

For the caterpillar embryos that survived the summer and the winter the obstacles seemed to increase rather than diminish. Some caterpillars never escaped the shell. Depending upon the vagaries of weather, which in a sudden turn could immobilize them in their tracks, sap their endurance, delay them until all their reserves of energy had been burned futilely, anywhere from three to thirty-eight percent died. Some even managed to crawl two-thirds of the way to freedom before collapsing. The laggards, those that emerged last at this time and those that had not even begun to hatch, were discriminated against by their fellows, and their fate was often death.

Once out of the shells the early risers rested; indeed many did nothing but rest for a whole hour. Over the next few hours each one stretched a bit, explored the immediate environs by thrusting its head about, and nibbled unenthusiastically at the spumaline. Whenever one did move, it spun before it in zigzag pattern an extremely fine silk thread, so fine in fact that it could be seen only with a powerful lens and even then only when the light glinted upon it. It was not long before the egg mass was webbed in silk. By this activity of the adventuresome the laggards were penalized because they now had to cut passages through shell and spumaline and silk. Some never succeeded.

It is an ill wind that blows no good, and tardiness is not an unalloyed misfortune. This is especially true of those individuals of a species whose lives straddle two seasons at a critical period in their development. Those that are in a state of rest, in an egg or a cocoon, those that are seeds or buds, those that are young tadpoles or nest-

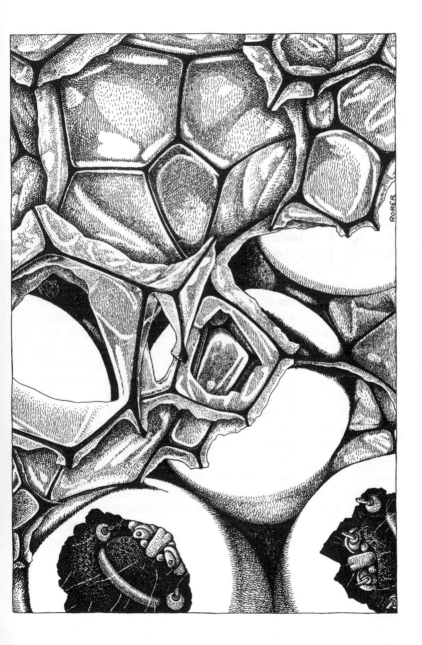

lings, and those that are in the vanguard of the great migrations, all are vulnerable to caprices of the weather. For them extremes of temperature, unremitting rain, or drought, can be catastrophes. In the case of the caterpillars, were all the eggs to hatch at once, and were a major catastrophe to strike, the entire colony would be wiped out at a stroke. As long as a caterpillar remains in the egg, however poor a shelter that may be, it can withstand some of the cruelest extremes of spring. Once it has left the shelter of the shell it is immensely vulnerable. Spring is surely a time of uncertainty, and the early plants and animals must be a hardy folk. If all of the first wave are struck down, however, those that have not rushed the season are alive to replace them.

This particular year was no exception to the rule of uncertainty. After nearly two weeks of luxuriant warmth the weather changed. Within twenty-four hours the temperature dropped from seventy to twenty. The frogs were silenced. A skin of ice locked the swamps, and three inches of snow fell. Crocuses disappeared. The venturesome honeybees, a delicate folk, froze to death. And what of the caterpillars? Those that had hatched huddled in a tight bunch on the sheltered underside of the egg mass. There they sat for the next ten days while the weather vacillated. Briefly the sun realized its spring duty and melted the snow, but a few days later a sleet storm beat savagely across the land. One by one the caterpillars relaxed their grip on the silken mat that they had spun in sunnier times. Some fell to the ground; others hung lifeless.

Survivors taken indoors at this time became active within minutes. Many were too weak to eat. Unhatched eggs rescued from the cold at the same time began to hatch by the following day. Compared to the rescued survivors they were a lively bunch. In the scramble for food they left little for the others. As the cold continued outside for four more days the death toll there rose. Not only had those caterpillars been too lethargic to crawl in search of food, there was no food to be had. The cold had also delayed the budding of the cherry trees.

Thus, when spring is too temperamental and too late, the early-hatching adventurers are annihilated. The future of the race then lies in the destinies of the conservatives.

V

And when they were up, they were up,
And when they were down, they were down,
And when they were only half way up,
They were neither up nor down.
A. Rackham—"The Noble Duke of York"

From some secret winter hermitage, a mourning cloak butterfly emerged and found its way to the stump of a gray birch that had been felled a few days earlier. The birch had been fractured in one of the ice storms and was beyond mending. Sap rose in abundance to the sawn surface. How could the cells of the root hairs know that the tree they were striving to nourish no longer existed, any more than does a heart that continues to pump blood, beyond all hope, to a severed artery.

The sap fermented in the beneficent warmth of sunlight. This winy odor attracted the mourning cloak whose appearance was a certain indication that spring had at last done with its tantalizing. One mourning cloak does not make a spring, but other signs offered confirmation. One had no farther to look than the stump. Although the root hairs were no longer pumping life to a tree, they were pumping it to a butterfly and to swarms of pale midges. Other invited guests included greenbottle and bluebottle flies, black blowflies that had permanently abandoned their winter crevices, lean gray flesh flies, hirsute tabanids, and a lively company of gray and of golden andrenid bees. These last are the sociable ones that do not live together in a single hive but have established individual subterranean nests in some sunny bank as members of a loosely knit community. They were continually interrupted by legions of Sepsidae, little ant-like flies, the males of which rushed about excitedly, tremulously waving their wings to attract females upon which

they immediately leapt. For these last at least the bacchanalia had developed to a saturnalia. Myriads of springtails also came and drowned happily in the intoxicating liquid. All these signs, and others, said that spring had arrived to stay; and, as if to add an exclamation point, a large bumble bee barged into the tippling throng.

By now some buds had not only begun to swell but were anticipating, by their yellow, red, and green suffusions, the advent of their bursting. That magic pigment, chlorophyl, by which plants gave the world its first solar energy cells, was just beginning to make its annual appearance in the deciduous forests. Fortuitous indeed was this appearance because those tent caterpillars that had survived the temporary setback of spring a week earlier had exhausted the last of their resources. A residuum of energy carried them to the tips of the cherries and apples. Even in this final expenditure of effort they played a continual Sir Walter Raleigh to themselves as they laid down silken paths upon which to tread.

Clad in velvety black these caterpillars measured a grand four millimeters overall, the space occupied by four letters on this page. As with all young creatures their heads were too large for their bodies, a feature responsible for the universal appeal of young animals and babies. The extremely delicate "hair" that clothed them was disproportionately long and also imparted an impression of cuteness, as down and fuzz so characteristically do. Aside from these disproportions they were undistinguished in appearance.

Infant though they were, learning to walk posed no problem. They hatched with that capability fully developed. As young inexperienced caterpillars they came completely equipped with the necessary skills. There is no evidence that they need to learn anything or indeed that they can.

If they did not take the time to spin silk, they could travel at the rate of one and one-half millimeters a second, a rate equivalent to about twenty-five times their own body length per minute. Whether this is reckoned as fast or slow depends on the point of view. Every observer tends to measure in terms of his or her own ability to cover distance. To the foraging chickadee the caterpillar's speed must appear slow. To a human eye it is excruciatingly slow and labored. A six-foot man walking briskly, which is about four miles

an hour, travels three thousand five hundred and twenty times his own length, that is, height, per hour or about fifty-nine times his own length each minute—unless, of course, he arrives at a mountain. The caterpillars, on the other hand, appeared to be moving at the same rate up-twig as down. If there was in fact any difference, it was so slight as to defy measurement.

Buds at the tips of twigs are normally "up" with respect to the twig and the rest of the plant; but this need not be so. A trip from the egg mass to a bud can be down as well as up, especially if the branch is gracefully arched and drooping. Apple trees even more than cherry trees stretch their terminal branches horizontally and downward. On the same branch one fork or another leads first toward the sky and then dips earthward. In any event since that which goes up must return down, the caterpillars are exposed to both of these contingencies and, therefore, are in a position to experience both sensations.

Realizing this fact and contemplating the synclines and anticlines of a thicket of twigs one begins to wonder if caterpillars can indeed experience up and down. Are they sensitive to gravity, and, if so, do they behave accordingly? Gravity is something that mankind tends to take for granted unless it is brought to his attention forcefully by a fall, or vicariously by reading of astronauts in space.

To the ordinary person, and perhaps even to the physicist, gravity is a mysterious force; nevertheless, it is a force with which nature must contend. The plant whose roots grew up instead of down would indeed be in difficulty, as would the bird that did not know it was falling when and if its wing-beat faltered, or the fish which, in fleeing from the osprey, swam up instead of down.

Obviously organisms can note the direction of gravitational force. They show it by their behavior. They reveal it simply by having and maintaining a characteristic posture. For human beings this normal posture is front end, that is, head, up. For most quadrupeds it is bottom side, belly, down. For the three-toed sloth lazing away its life in tropical jungles it is back side down. For that curious bug of the ponds, the backswimmer, it is also back side down. For the sleeping bat it is head down. For plants it is roots down, stems up.

Responses to gravity do not end here. There is, for all, the ever-present problem of indulging in sensible coordinated movements in a field of force where everything naturally falls down. As is always the case in these matters, nature has a varied bag of tricks for solving each problem. Human beings are endowed with a complete set designed to provide alternatives should any one system fail. Such redundancy is an indication of the importance of maintaining proper gravitational relations. To this end the inner ear, the muscle senses, the soles of the feet, and vision all play a crucial role.

Rattling around in the inner ear are small ear stones which the pull of gravity causes to press against certain sensory hairs. Since each hair is "wired" differently to the brain, the brain knows when a top hair or a bottom hair is being pressed, hence, whether the person is rightside up or upside down. Even lowly jellyfish and crayfish are similarly endowed, but insects must rely on other devices.

Since gravity causes things to fall, different parts of the body, limbs, organs, and masses of muscle, also tend to fall, thus creating tensions and pressures. A device for monitoring these events could convey a great deal of information about the direction of gravity. It is hardly surprising, therefore, to find, in vertebrates, sense organs embedded in muscles and tendons. These inform the brain of the pulls and pushes as the body changes its position, and the brain, knowing the source of each message, can assess the ups and downs of the body as a whole. The average person never appreciates the tireless attention to duty of these muscle senses until he or she is denied their information, a situation that occasionally ensues when a cramped limb "goes to sleep." When that happens, the brain has no idea where the limb is until the eye locates it.

In caterpillars, or in any insect, you can search in vain for the muscle senses. They simply do not exist. Upon further reflection, the omission should not be surprising since there is no elaborate internal skeleton upon which to hang the muscles. Instead insects have their muscles attached to an external skeleton, the tough articulated cuticle.

The cuticle absorbs all stresses and strains. What better place to locate sense organs than there? And there they are, small oval and elliptical membranes whose distortions are signaled to the brain.

When the caterpillar is crawling head down, different areas of the cuticle are strained then if it were traveling head up. The brain for all its minute size "knows" the topographical source of the signals, hence, up from down.

There are even more precise ways of accomplishing the same end. Wherever two joints come together, the cuticle is equipped with fields of stiff little spines. These are rubbed by the joints in a symmetrical manner if the caterpillar is in a normal position and posture. Any change, a turning of the head, an extension of the leg, a pulling of one side of the body by gravity, stimulates the hairs differently.

Nature has endowed the young caterpillars with all of this exquisite equipment. Do they, however, behave differently with respect to up and down? Apparently direction at this time of life makes little difference. Caterpillars are equally at home going one way or the other. They show no preferred direction, accepting the irregularities of their arboreal highways as they find them. Other animals around them may have preferences, the brown creeper working its industrious way up the trunk of a nearby pine tree and the white-breasted nuthatch working just as industriously down head first on the trunk of a maple. Caterpillars are directed by other exigencies that prevail over gravity. None of these observations means, however, that caterpillars do not sense gravity. Because an animal appears in its behavior to be oblivious of events and forces around it is no indication that it is insensitive to them.

A sense of immense importance is touch. Clearly caterpillars know when their feet are on the ground, whether the ground is up or down. Human beings are informed of their being properly upright on the ground by pressure senses in the soles of their feet. Caterpillars too are informed by senses in the feet. If the feet lose touch, the brain assumes that the body is upside down and must be turned over.

Righting reflex comes as close to being a universal behavioral response as any. No one who has ever observed a turtle turned turtle, straining its scrawny neck to extraordinary lengths to lever itself over, can doubt the importance of being rightside up. Animals will struggle to exhaustion to regain their feet. So important is correct

position that many animals have special means of achieving it. The click beetle has a hook and catch mechanism at the joint between its abdomen and thorax which it can engage when it is on its back and which, when the beetle straightens its body, disengages with a snap. Flung into the air the beetle with luck lands on its feet. Many winged insects, finding themselves on their backs or with their feet no longer touching anything, simply begin flying. The caterpillars, being wingless but supremely flexible, manage very well by squirming.

The world in which the caterpillars found themselves on this spring day was in every sense their own. It was *their* perceived world to which they had to adjust. It was a small world. True, the destinies and characteristics of their little cosmos were influenced by events of immense magnitude. Winds curling over the mountain ridges or sweeping across the fields buffeted the bushes so that the world was one of motion. Presumably the caterpillars were immune to motion sickness, vertigo, and sudden changes in acceleration. Theirs was not terra firma. Once having left the flaccid existence of the egg they could no longer enjoy relaxation while they remained caterpillars. Not until they passed into the life of pupae could they stop standing and clinging. To relax at any time before this would be to court disaster because then they would be cast into a world where their perceptions would serve them ill.

This greater world of spring was rich in features that the caterpillars could never appreciate. A vast cosmos of form and color existed beyond their capacity to sense. Across a sky of robin's-egg blue marched legions of dazzling white cumulus clouds, rising from behind mountains and disappearing over purple hills. The horizon itself undulated in rolls of hills where dark green clumps of white pines gave definition and depth to the softer roseate background of maples and where patches of birches and beeches accentuated the contrast with dashes of light. And in the middle distance the brown of last summer's grasses, sedges, and forbs accentuated the fresh green of new meadow grass. Here and there the sky looked at itself in small ponds and swamps. None of this could the caterpillars see. Nor could they even see the form of the cherry bush on which they lived. They could barely see each other.

Eyes they have, but these are suited to their own needs. Located on what in other animals would be called jowls are six little buttons arranged in an imperfect semicircle. It seems an odd location for eyes until one considers the field of vision that best serves a caterpillar's mode of life. Only a vision of the twig to be trod and the leaf to be eaten is important.

It is said truly that man is the one creature that walks perfectly upright, holds his head high, and contemplates the heavens. His eyes are placed in accordance with his needs: together, in the front of his face, and toward the top. Few other creatures look into the sky, probe the distances, and scan the constellations. Birds, it is true, employ the stars for navigation, but they are creatures of the sky and are in the sky. The eagle has keener vision than man, but its gaze is earthward. For most animals, things of interest are earthbound—enemies, prey, food, trail marks, nests, and mates.

Caterpillars too are earthbound. To the small, small things are important, therefore caterpillars are also extremely nearsighted. Additionally, their vision is astigmatic and blurred despite the fact that each small eye is equipped with two reasonably good lenses. It is even possible to see firsthand how the visual world appears to them by dissecting the eyes (ocelli) and photographing through the combined lenses. It is a fuzzy but recognizable small world that blurs off into nothingness beyond a distance of two to three centimeters. It is no wonder that a hungry caterpillar can pass within a few centimeters of a cherry leaf and never see it.

Form is a different matter, provided an object is vertical, contrasty, and of the correct width. The angle that the limits of an object make with the eye is critical. If a cherry twig is placed upright in the middle of a flat surface in which young tent caterpillars are crawling, they see it only when they are a few centimeters away because only then does the twig loom large enough on the tiny retinas to stimulate. What is important is the boundary, the edge of the twig against its background. Just as a person in an unfamiliar dark room moves cautiously forward with extended arms moving back and forth, so do the caterpillars periodically hesitate, raise their foreparts, and wave blindly from side to side. In this manner they discover by contact objects they have not seen. It is equally impor-

tant that this waving movement causes the image of the boundary to pass successively over all six ocelli. Scanning compensates for the small number of ocelli and the absence of a retina with thousands of cells. Even then the picture obtained is like that obtained on extremely coarse-grained film. The coarseness can be imagined by viewing a newspaper photograph through a magnifying glass.

Despite the primitive nature of their vision, caterpillars, especially in later life, probably make use of it in identifying stems as things to be climbed. The ocelli also provide their owners with color vision—a sense that is usually taken for granted. In the entire animal kingdon only a small minority are blessed with a sense of color; the rest are colorblind. Among the fortunate are: insects, bony fishes, birds, some reptiles, and primates. Not even all of these see the same world of color that man does. Insects have their own primary colors, not the red, green, and blue of primates. In that part of the light spectrum that is invisible to mankind, the ultraviolet, insects see color. On this spring day, the white crocuses, the white magnolia, the white snowdrop, and the white house might all appear differently to the bumblebee. If one reflects ultraviolet and the other absorbs it, the bee sees the whites as two different colors.

The caterpillars on their first exploration of the cherry twig are crawling along twig highways that stretch indistinctly ahead of them for a few millimeters and fade away in a fuzziness that does not even provide them with a horizon. They do not see when they have arrived at an end until suddenly there is no footing ahead. Briefly they stretch and probe blindly into the void, until, finding no firm path, they turn back and retrace their steps.

This spring day, their first extended venture into the world, ended much as would most days of their short lives. It was a warm featureless day of fuzzy light and color, with an up and down that did not really matter.

VI

I 'spect I grow'd
Harriet Beecher Stowe—*Uncle Tom's Cabin*

The period from spring through early summer is a time of paradox in the appearance of the countryside. At one and the same time it becomes softer yet more substanial. At the beginning of the season features are more boldly delineated, especially where the hills present sharp profiles against the sky and the trees stand in fine linear outline. What in earliest spring appears as an etching now acquires the diffuseness and blending of a water color. Where once it was possible to see through the skeletal coppices to the distant land and to look past the gray maple saplings into the deeper forest, a green curtain has now descended. The new leaves that color and soften the harsher contours curiously enough impart a new dimension of solidity to the forest. Fields too are transformed from thin, threadbare, two-dimensional coverlets to thick soft blankets. The secret is that living stuff is increasing in mass. There is more of it. It begins to fill all available space.

Irresistible forces are at work, forces that had been held in check only by winter's low temperatures. Released by summer's warmth, as a giant from his chains, these forces are constructing, from the ethereal substance of the atmosphere and the formless humours of the soil, living stuff of formidable mass. Where all else in the world is running downhill toward increasing disorder, life is building. Against all destructive forces it builds toward order. Inexorably it keeps adding to itself. It is growing. It is growing as no crystal grows and as neither mountain range nor river delta grows. Life is growing by building from primal elements improbable molecules, linking these together in even more improbable combinations, and mold-

ing the whole into cells which by multiplying become leaves, flowers, bark, skin, muscle, and nerves.

While the essence of growth is increase in mass, the quality of growth is design toward form. All growing parts must maintain proper relationships with one another. In the absence of design there are calluses, cancers, and chimeras. Order forbids the reality of basilisks, centaurs, gryphons, and sphinxes.

The sheer immensity and complexity of the process of growth would suggest little latitude for versatility, but this is not the case. Hundreds of thousands of variations on the theme reveal themselves in the multitudes of species that populate the land, the streams and ponds, and the oceans. Each solves the problem of growing in its own characteristic manner. Disharmony and death alone halt the relentless thrust of life to overwhelm the planet.

Grass, herbs, and trees grow at their tips, of stems and roots alike, and add to their girth without restraint. Bark, aged and cracked, sloughs away as it is replaced from within by the expanding trunk. The power of growth shows itself in the gnarled boundary trees where living tissue grows around and over the barbed wire that had been stapled to the smooth young trunks a generation earlier. The only remnant of the rusted-away fence is the segment of wire now deep within the trunk.

The soft-skinned spring lambs, the birds, the children in the village, indeed all animals with internal skeletons, add length and girth to their bones, and their supple skins and muscles stretch and grow in harmony to accommodate the skeleton. As growth proceeds from within it meets no restraint. Cells of the skin die at the surface as they wear out and are pushed from within. Constant replacement of cells that slough off insures that the increasing mass of the body suffers no restraint.

The tent caterpillars entered the world with a constraint not imposed upon plants and vertebrates. It was their destiny to be encased in a hard, non-living coat of armor. They are not alone in the possession of this unyielding external skeleton. It is an incubus they share with crabs, lobsters, spiders, centipedes, and an almost endless list of relatives.

Imprisonment within an external skeleton means that its owner

can grow only to the extent that the skeleton can stretch. Beyond that point the animal is crushed within itself unless room can be made. In this respect it resembles a growing boy who outgrows his clothes and can find comfort only by replacing his too-tight shirt and trousers with larger new ones. The caterpillar can find comfort only by growing a larger, expandable skeleton, folded and packed within the old until needed. At time of need the old must burst and be cast aside. This is a time of great danger. If the machinery of molting goes awry, slow strangling death ensues.

Molting is a trick worthy of any magican. The concatenation of events leading to the production of new skins and the discarding of old occurs six times in the lifetime of the tent caterpillar. It is a process orchestrated by a trio of hormones that are produced in a carefully programmed sequence. The first performance comes a scant seven days after the caterpillars have hatched from the eggs. For some colonies, especially those feeding on cherries growing on the southerly borders of roads, in hedgerows along sheltered, sunny undulations in the fields, and those upon apple trees close by houses, the molting day arrives early. These are the colonies that will mature while others have yet to experience their first molt. In human experience such a degree of slow development as the latter exhibit would be viewed with alarm as retardation. As was the case with hatching, however, success does not always go to the quick. In nature conformity and allegiance to unrelieved monotony is folly. Only mankind strives relentlessly for the phantom perfection of uniformity.

Each tent caterpillar colony is different, and each colonist an individual. Despite the individuality, however, all are the playthings of forces beyond their control—forces from the mundane to the celestial. Responding almost mechanically to the same internal and external cues day in, day out, they fed at the same time, ate at the same rate, consumed similar amounts, and extracted the same nutrients from the cherry. But, as in small ripples upon an ocean wave, there are subtle differences superimposed on the cyclic swells of behavior. Some caterpillars feasted by sheer luck on more succulent leaves than others. Some wasted more time traveling from tent to twigs. Some had more of a bent for exploration. All

sorts of delays, interruptions, and misadventures prevented all members of the colony from being smoothly, socially rhythmic, from being rigidly conformist automatons.

Nevertheless, at the end of the first week of independent life, there came to each a point when the infant skin, the skeleton that had been acquired in the egg, could stretch no farther. Only a few hours separated this advent of critical repletion in the most advanced individual from its appearance in the most retarded. Thus, in the span of a few days the colony was transformed from a devouring army into a collection of lethargic, stuffed gluttons. Within the latitude allowed by variation molting was a social event.

For many individuals hours of preparation for the first molt had already been going on within the brain. In no way was this a mental preparation, a nerving for a hitherto unexperienced ordeal. It was not even neural. Instead, an insignificantly small group of cells in the midbrain had been manufacturing a special hormone. They manufactured it, and they waited. As in so many of the crucial activities of caterpillars the behavior of these cells is tied to events of a far greater world than that of the tent in the cherry tree.

It comes as no surprise that the movements of the earth in space influence events of great magnitude and import, but nothing under the sun is too insignificant to take advantage of astronomical events. The timing of egg development was no exception. The cells of the caterpillar's brain are no exception. They live to the tick of their biological clock, the hands of which are set to day and night. If the tent caterpillars are at all like other caterpillars, and there is little reason to doubt this, the advent of sundown places the cells on alert. For ten hours they wait for a signal to release their hormone. If at the end of that time the signal has not been received, the cells become unheeding. Not until the same time the following day do they resume their vigil. It is a marvel of timing that they can measure off a span of ten hours and an interval of twenty-four hours between the beginnings of each ten.

The signal that they await is as astounding as the timing of their vigil. It is nothing more intricate than weight. If the caterpillar has achieved or exceeded a critical weight by the time the brain cells have become alerted, the hormone is released, and the caterpillar

loses its appetite. If the caterpillar is underweight, the hormone is retained, and at the next approaching mealtime the caterpillar sallies forth to feast. But, simple as the signal appears, the sense by which the caterpillar weighs itself defies explanation.

The first caterpillar to reach the critical weight in the colony of which we have been speaking suddenly lost its appetite just after the sun had set. Its brain cells had relieved themselves of their burden of hormone. The hormone flowed down to a set of glands located in the first segment of the thorax, whereupon these cells released into the blood a hormone of their own. This tide, of the hormone ecdysone, bathed the nervous system and all the other organs of the body as well. It was this that killed the appetite and brought on the great lethargy.

None of these momentous internal happenings was visible from the outside. Some caterpillars still gnawed at the cherry leaves, while a red-eyed vireo picked off a few caterpillars, and ants carried off others. The quiescent caterpillar remained back at the tent where it lay packed side by side with others that had reached the same weight. Inside the inert bodies paroxysms of activity were occurring. Tissues were being destroyed, redesigned, renovated, all under the power of the hormone ecdysone. It caused the epidermis to pull back from the old tight cuticle, fluid to be pumped into the intervening space, and layers of new cuticle to be formed by the cells of the epidermis. The fluid digested the inner layers of the old cuticle but spared the new.

Although ecdysone was a prime mover evoking hitherto unrealized potential from cells and tissues, it could only call forth that of which the cells were inherently capable. Even then its call was selective. Had the cells heard only this siren, they would have then and there begun fulfilling their ultimate destiny of producing a moth. As it was, there is a time and season for everything. This was neither the time nor the season for taking to wing. Premature coming of age, of any of the life forms comprising nature, would throw the whole intricate ecology into chaos, destroying not only the miscreant who was out of step but others as well.

To forestall undue hastening of adulthood a third hormone is provided in the tent caterpillar. This hormone, acting the part of

supervisory wisdom, insures that the cells of the epidermis form only caterpillar cuticle. So instead of building moth scales they formed hairs, as are proper to a caterpillar. The pattern was somewhat more complex than the one it was replacing, the hairs longer, the colors a bit more daring. With each succeeding molt the colors would be enriched until the caterpillar finally attained its full resplendent dress. It would still remain a caterpillar. Ponce de Leon may not have found the fountain of youth, but the caterpillars, like all other insects, had discovered the hormone of youth. If they continued manufacturing juvenile hormone, they would remain young forever. Indeed, in the laboratory it is possible to prolong youth by injecting juvenile hormone into insects that are approaching the end of their adolescence. The insects continue to grow; they grow old, but they never grow up. The thought of immortality is seductive, but there is something revolting at the thought of giant, aged infants, whether they be caterpillars or human beings.

With all the preparations completed the caterpillar had only to escape from its prison. Perhaps another hormone signaled the time. In any event at some ordained moment the caterpillar shook off its lethargy. It began a series of violent muscular contractions. At the same time the volume of its blood had increased, providing an internal pressure that could be localized in one part of the body or other by controlled pumping. Subjected to all of these forces, the old skin, thinned and weakened by the digestive fluids that the epidermis had pumped forth, split at the neck and back just behind the head. The caterpillar bulged forth from the rent. The old head skeleton remained for a while as a mask over the face, but the legs were free, and sufficiently firm to grasp the silk of the tent. With waves of convulsive heaves the caterpillar crawled out of its old skin.

Fortunate indeed for this caterpillar that its transformation occurred at night. At no other time was it quite so vulnerable to attack as when freshly molted. The new skin, while comfortably roomy and flexible, was soft. It provided no protection against the jaws of predators and the stings of parasites. In fact the caterpillar was even vulnerable to attack by cannibalistic members of its own

species. Added to these handicaps was a pathetic feebleness of locomotion. Not until the new skin was stretched and hardened to its new dimensions could the caterpillar venture forth to feed.

During the night still another hormone worked its biochemical magic and provided for the hardening, tanning, coloring, and waterproofing of the cuticle. The same sequences were taking place in the bodies of all members of the colony, some slightly earlier, some a bit later. By morning the colony presented to the eye a picture of small, uncomfortably tight, rather dark caterpillars intermingled with larger, fuzzier, lighter, flabbier individuals. The former huddled motionless; the latter appeared more restless. As the sun rose, signaling mealtime, the restless ones, once again a ravening horde, began marching in line to the tips of the cherry twigs.

Growth had been almost as traumatic an experience as birth. During this period of travail for the caterpillars the tissues of the cherry had been growing, evenly, smoothly, as continuously as the flow of time. Across the fields, in a warren tunneled beneath a mound of wild raspberries, a family of newly born cottontail rabbits were growing as imperceptibly and uninterruptedly as the intrusive advance of a flood tide. No growing pains for them. Nor were the young gray squirrels in the drey in the red oak struggling through molts either. Growing for all of these creatures was a play that unfolded evenly from birth to death with no curtains to measure the scenes and acts. Could it be just an accident that those things whose physiology ordains that growth flow as a stately processional are among the long-lived species, while few among the molting creatures ever become the venerable of the earth? So much planning, so much effort, and so much complexity for lives so transient!

VII

*Often on a work of grave purpose
and high promises is tacked a purple
patch or two to give an effect of colour.*
Horace—*Ars Poetica*

So many things are seen to their best advantage from middle ground. But for each separate thing middle ground is a unique point in space, as are also foreground and distance. With a forest, distance is a matter of miles—from which situation it is seen as through an imperfect lens and as an uncertain line fuzzing the clear profile of the mountain. At the other extreme, it is equally indistinct and elusive when you are upon it. When you are within it, it cannot be seen for the trees. Only from middle ground is its essence as a forest revealed to the fullest.

So it is with all things. Yet, to be appreciated fully for what they are, they must be viewed from that particular distance which neither reduces them to a fragment of some greater whole nor reduces them to lesser fragments of which they are the whole. This phenomenon is equally true of forests and caterpillars.

Relative to a caterpillar, distance is a matter of yards. Within this frame a tent caterpillar is a thing singularly lacking in beauty unless it is in the eye of the yellow-billed cuckoo who views it with the critical air of a gourmet. Appetite is a rose-colored spectacle that imparts to objects seen a beauty that is no less aesthetic merely because it is realized by so gross a system as the gut. To our eyes the caterpillar at this distance is as unresolved as the forest on the shoulder of the mountain. Its form is indistinct, its color somber, indescribable, and false. At a distance the true colors of things are concealed. The purple hills on the horizon are not purple when you climb them. The lake is not blue when you swim in it. The tent cat-

erpillar is not a nondescript hue of vagueness when you approach it.

From middle ground the tent caterpillar, once it has passed its early molts, reveals an unsuspected beauty to the pure and unbiased mind. Abandon all your preconceptions and prejudices. Do not think of ravaged trees. Do not be revolted by a mode of locomotion that errs only in being different from your own. Focus your eye from the middle distance, from the vantage point of a few inches, and see. Look beneath the furry coat as you would strive to see the man beneath his beard.

Extending down the middle of the back from head to tail is a strikingly white line, bowing slightly at each segment and constricted just enough at each intersegmental junction to give the whole stripe the appearance of a standing wave. On each segment there is a pure baby-blue spot to either side of this stripe. Below these on the sides is a mottled wash of blue and gray. The arresting features of these colors are the whiteness of the white and the limpid atmospheric impression of the blue. Yet, there is no gloss here, no iridescence. The colors are flat. There are in fact no pigments.

As every artist realizes, the colors of things undergo startling transformations as daylight changes and as an object is viewed from different angles. If the white stripe and blue spots change at all with angle of viewing and changing of daylight, the change is so subtle as to escape notice. From all angles, in the high sun of noon and the yellowing sunlight of afternoon, the white and blue remain true to their colors. Wetness is another matter. Many objects reveal their colors only when they are wet. Pebbles on a rocky seashore lying beneath calm water of an ebbing tide flaunt marvelous stripes and swirls and mottles of grays, whites, blacks, reds, and greens. As the tide ebbs these same pebbles drying on the beach fade to anonymity. With the caterpillar quite the reverse is the case. Immersed in water the midstripe fades, and the blue assumes a greenish cast. Herein is provided a clue to the colors.

In nature caterpillars are never exposed to glycerol, except to that which they themselves provide before their winter ordeal. If, however, the mature caterpillar is immersed in glycerol, the midstripe becomes translucent and nearly disappears. In other solvents it becomes transparent. The blue spots fade to an undistinguished dark

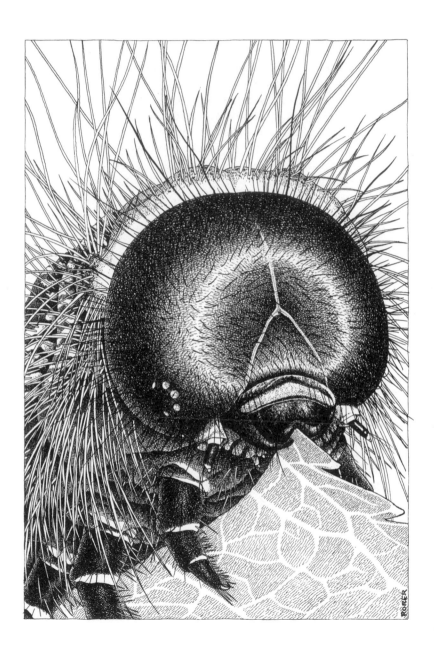

greenish blue. The beauty vanishes. If you analyze beauty, whether physically or intellectually, you often destroy it. With the caterpillars we have now intruded beyond the middle ground and invaded the foreground, searching for understanding.

The cliché that "beauty is only skin deep" is, as might be expected, too all-encompassing to be true. There are neither white pigments nor blue pigments in the skin of the caterpillar. Nor, for that matter, are there white or blue pigments beneath the skin. Both observed colors result from the white daylight being broken up and scattered by minute transparent particles. The areas of cuticle in the regions of the blue spots and white stripe differ from other areas of the integument in bearing microsculptured transparent filaments. These are of the dimensions of the wave lengths of blue light. When the white daylight, consisting of a mixture of all wave lengths, strikes this microscopic pile, more of blue than of other wave lengths is scattered. Other wave lengths pass through the mat of filaments to a layer of black pigment lying immediately beneath, where they are absorbed. Only the blue bounces back to the eye of the observer. The same scattering phenomenon occurs at the white stripes, with one important difference. Beneath the stripe there is no absorbing layer of black pigment. Thus, although light is scattered with a preferential bouncing back of blue, the remaining light, passing into the body, is reflected back out through the filaments and so dilutes the scattered blue that it is seen as white.

The tent caterpillar is one of the few insects possessing blue color produced by structural means, the blue known as Tyndall blue. Dragonflies are competitors for honors, but employ somewhat different systems. The numerous other insects that sport structural colors are iridescent and achieve their effects, not by selectively scattering blue wave lengths, but by diffracting light into its many wave lengths as do prisms and oil droplets.

Of what use these colors are to the tent caterpillar we cannot fathom. A near relative, the forest tent caterpillar, employs the same cuticular tricks to generate colors. Instead of a midstripe it has a line of keyhole-shaped spots, the centers of which are white.

Along each side there is a continuous blue stripe, the same Tyndall blue.

Perhaps greater understanding does bring appreciation, and perhaps appreciation does reveal cryptic beauty. Perhaps there are two levels of beauty, the one elusive, instinctive, emotional, the other revealed only by analysis. Analysis, however, reveals beauty only when the parts are reassembled and the whole is viewed from the middle ground of its best advantage.

VIII

The distance is nothing;
it is only the first step that is difficult.
Marquise du Deffand

If there is any single feature that characterizes animals, it is the ability to move. Yet even that characteristic is an ambiguous criterion because there are plants that move and animals that do not. Indeed, it is not that plants fail to move, but that they move in their own time scale. In accelerated time roots burrow through the soil at a frenzied pace, grass stolons no longer creep across the ground but outcrawl the earthworm, and climbing vines twist, coil, and strangle with reptilian speed. Yes, plants are only slower. Nor can they stand accused of being rooted to the spot, because slime, molds, and colonial algae give lie to that accusation when it is made as a generalization. By the same token, the barnacle, the sponge, the coral, and the scale insect are as rooted as any plant.

Nevertheless, motion in our time scale is an animal trait. The ability to move from place to place, not as a passive plaything of the winds and ocean currents but under one's own power, is to be an animal. To travel is to court danger, but to be able to travel is to be free.

With the advance of spring the whole animal world was on the move proclaiming its independence by its mobility. During the night a warm gentle rain had soaked through grass and leaf litter, eventually flooding the burrows of earthworms. Legless and lowly, the worms nonetheless set out on long journeys. Morning caught hundreds of them stranded on paved sidewalks and hard roads. Here their ancient way of crawling, so effective in friendly soil, failed them. The waves of muscular contraction passing rhythmi-

cally from front to back could pull the body forward only so long as there was a rough surface against which to gain purchase.

Other worms, frail coils, lashed themselves through the placid water of the barnyard horse trough. Ancient superstition taught that these thin creatures were horsehairs that had come to life in the rich brew which accumulated in troughs, puddles, and quiet pools. Science in erasing the mystery has impoverished the folklore but has embellished the complexity of the face of nature by revealing a complicated life history. The horsehair worms spend part of their lives as parasites in grasshoppers, from whose bodies they must eventually escape to find their way to standing waters. Without the assistance of rains and freshets, their feeble locomotion would certainly be unable to do the task.

The lashing mode of progression, so common among aquatic creatures from marine worms to mosquito larvae, appears strikingly inefficient, yet obviously serves these creatures well. Some, like the larvae of midges, drive themselves forward through the water by lateral undulations passing from head to tail. Mosquito larvae thrash their tails alternately left and right to pull themselves tail first through the water. There are almost as many modes of progression as there are kinds of animals.

Where the brook emerging from the forest formed a small pond at one corner of the lower meadow, hundreds of other animals walked, crawled, rowed, skated, undulated, or flew. Swallows from the barn skimmed acrobatically over the surface. A mixed herd of Holsteins and Guernseys came in their slow quadruped fashion to the edge to drink. On the face of the pond waterstriders rowed across the surface film, minute springtails leaped with powerful thrusts of their tails against the surface tension, and whirligig beetles spun dizzily with the frenzy of inebriated dancers. Beneath the film the waterboatmen and backswimmers rowed themselves to the surface to breathe, then back to the bottom to feed. Diving beetles paddled with hair-fringed legs to the surface where they captured bubbles of air. With these diving bells they thrust to the bottom to remain as long as the air supply permitted. On the bottom itself myriads of aquanauts were swimming, walking, crawling, undulating—some slowly, deliberately, laboriously,

reluctantly, others with remarkable bursts of speed. Nymphs of dragonflies, anticipating the jet-age by millions of years, forced powerful streams of water out of the anus when threatened by a predator. They often achieved short bursts of speed up to 250 cm/sec, exceeding even the whirligig beetles whose top speed in bursts is 100 cm/sec.

Every body of water, flowing or standing, boasted its active population, each member with its own mode of locomotion and gait. Even the stagnant water in the gutters of the house provided opportunities for the expression of all techniques of progression. Microscopically small waterbears, those enigmatic animals whose place in the scheme of things is a mystery, crawled with eight stubby legs in the rich bottom sediment. Amoebae, mindless gobs of jelly flowing in all directions at once, together with paramecia whose waves of beating cilia spiraled them forward and backward with the blindness of mechanical toys, populated a still smaller world. And everywhere other one-celled animals whipped themselves along with a single long flagellum.

In the aerial world where the heavy fluid drag of ponds and puddles did not exist, motion reigned supreme. Only the choice of habitat, bodily design, and energetics imposed limits. In the air itself creatures whose ancestors had parted company so long ago that kinship was forgotten flapped, rowed, glided, and soared. Among the nectar-rich profusion of meadow and roadside flowers cabbage butterflies, clouded sulfurs, checkerspots, and blues flapped along as casually as boulevardiers. More businesslike bees bustled with the single-mindedness of impatient shoppers while erratic skippers added a frenzy to the scene by their dashing flight.

Over the pond the summer's first dragonflies hawked for mosquitoes, midges, or any weak flying insects that had not already fallen prey to the swallows. A sparrow hawk hung on fluttering wings over a patch of tall grass that almost but not quite concealed a meadow mouse. At the very top of the sky a vulture rode the thermals on motionless wings.

Earthbound creatures also were free of the tyranny of the drag of water, but, compelled to move on solid surfaces, they had both to overcome and to take advantage of friction. While swimmers and

flyers progressed by pushing against elastic and compressible substances, and consequently displaced large volumes as they moved, land-dwellers pushed against an unyielding ground. And if by chance the surface moved, as in loose sand, or became nearly frictionless, as when ice glazed the ground, they lost their footing. The cows, the farmer's dog, the meadow mice and rabbits, the squirrels in trees, the farmer himself, walked, waddled, ran, loped, bounded, jumped, leaped, trotted, paced, or galloped. Each had a gait or choice of gaits suited to body size, center of gravity, weight, and speed; but for each the problem was the same: to push against the ground so as to move the center of gravity of the body forward, to balance on the pushing legs while other legs were advanced into a more forward position.

Not the least of the problems that terrestrial animals encounter is that of supporting their heavy bellies. Vertebrates have solved this problem by slinging the belly from a rigid backbone that is supported at four corners by legs or balanced upright. Animals that rest completely on the ground, as do worms and snakes, have no concern with support; their bellies drag. For life in bushes and trees, however, on twigs, stems, and leaves, the problem of balancing a big belly in the absence of a backbone becomes acute.

Caterpillars are mostly belly. They have relatively small heads followed by three segments, each bearing a pair of legs, and these in turn are followed by ten segments of abdomen. Nearly eighty percent of the body is abdomen. The six weak legs, located far forward, are incapable of balancing such a huge abdomen on a leaf or stem and cannot retain their hold if the abdomen sags to one side.

A solution to the problem was found when the ancestors of modern caterpillars developed props for their long, heavy abdomens. Tent caterpillars, like most, are equipped with four pairs of props midships and one terminal anchoring pair. There is nothing particularly spectacular about these prolegs except that they are not true legs and are useless for walking. Each is merely a hollow extension of the body and, like the body, kept rigid by blood pressure. Each is crowned with a row of minute hooks for grasping. In the normal or relaxed position the prolegs are extended, and the crochets passively hooked onto the leaf, twig, or silk trail. The only active move-

ment of which a proleg is capable is that of unhooking the crochets by contracting a small muscle. Simultaneously the proleg is lifted by action of the abdomen. But neither with its legs nor its prolegs can the average caterpillar walk. How then does it crawl with such agility along twigs, and how can ground-loving species like woolly bears run with such remarkable speed?

Progression is possible because the abdomen is segmented, each segment has its own set of muscles, and the contraction and relaxation of the muscles in all the segments are beautifully coordinated. The first step, which is really not a step, begins with a movement of the extraordinarily elaborate musculature of the body. The tail is raised, bent forward, and lowered again. This action places the last pair of prolegs in a slightly advanced position but leaves the back humped. With the last prolegs tightly anchored the caterpillar contracts muscles in the next forward segment to straighten the hump. This of course creates a new hump still farther forward. As each segment contracts, beginning in the rear, it expands the segment ahead of it. The tuck runs forward till it pushes the head a short distance along the twig. Simultaneously, as the segments are being tucked and untucked, other body muscles raise each pair of prolegs at just the right moment so that they are carried forward to a new position. The whole coordinated movement resembles a traveling wave running from back to front, pushing the head forward. The last pair of prolegs then again changes position, and the whole process is repeated. The speed of crawling is determined by the rapidity with which the movement of each segment can be accomplished and coordination of all segments achieved. Execution of a rapidly flowing movement is analogous to an accomplished woodwind player having the dexterity and coordination of fingers to play a smooth arpeggio.

Insofar as the caterpillar fraternity is concerned, tent caterpillars represent one of many standard models. Some species have improved upon the basic mechanism or adapted it to their own peculiar style of life. The hustling woolly bears that are seen racing about in the fall have very short prolegs and fewer crochets than strictly climbing caterpillars.

Inchworms are the most innovative species. They are among the

speediest of caterpillars, having streamlined their crawling machinery. They have eliminated three middle pairs of prolegs. When
the last two pairs are moved forward, they are brought all the way
to the front of the body where the true legs are located. This action
throws the body into a large loop. Now the front legs release their
hold, the body is straightened and pushed forward a distance equivalent to five body segments, the front legs secure a new hold, and
the process is repeated with as much leisure or haste as the caterpillar deems necessary.

In the garden, especially if you grow cabbages, cauliflower,
or broccoli, you can find other caterpillars that have adopted
the same technique of crawling as did the inchworms, but have
never exploited it to its fullest. These loopers have one more
set of prolegs than do inchworms. As a consequence their loops
are less contorted.

Other species employ speed only in dire emergencies. Leaf-rollers, as their name implies, live in tubes made by rolling leaves
and binding the rolls with silk. Normally they are capable of crawling with commendable speed. If, however, a tubular home is torn
asunder by a predator, the caterpillar can achieve lightning speed
in escaping by vigorous lateral thrashing.

Some caterpillars have sacrificed speed to the demands of bizarre
modes of living. This is true of those smallest of caterpillars, the
leaf-miners, that spend their entire larval lives in the two-dimensional world between the upper and lower surfaces of leaves. They
themselves are almost two-dimensional, being greatly flattened to
accommodate to their flat world where crawling is quite restricted.

If only something could be done with the heavy dragging abdomen it might be possible for a caterpillar to walk in a proper manner. A few caterpillars, quite unrelated, have managed this feat,
and they have done so by building close-fitting cases to support,
protect, and conceal their abdomens. The bag-worms spin bags of
pure silk or of silk ornamented with fragments of leaves, twigs,
grass, grains of sand, excrement, or even minute molluscan shells.
The pistol case-bearer constructs from silk and the pubescence of
leaves a case shaped like an old-fashioned pistol. As a consequence
of having their abdomens completely incased so that they cannot

assist in locomotion, the bag-worms and case-bearers rely upon their true legs. With only their heads and legs projecting from the cases they walk about quite respectably, carrying their homes with them.

However much one may disparage the stolid abdominal crawling behavior of caterpillars, the fact remains that they possess probably the most versatile abdomens in the animal kingdom. What other animal can stand on the bottom of its abdomen and extend its entire body rigidly into space to resemble a twig, as some inch-worms do? What other animal can stand on its abdomen and swing its body in a 360 degree arc? Because of the strength and flexibility of its abdomen, a caterpillar can reach out to objects beyond the span of its short legs. When it arrives at the end of a twig or has eaten a leaf completely, it swings in wide searching movements for a new leaf or new foothold. When the tent caterpillar spins its tent, it relies on the versatility of its abdomen to enable it to secure silk threads to one spot and extend them to a distant one. The abdomen is to the caterpillar what arms are to a person.

It is all the more remarkable that there is no rigid, articulated skeleton to support all of these activities. Rigidity and form are maintained solely by the pressure of the blood circulating freely without benefit of blood vessels inside the tough muscular body wall. If a caterpillar loses blood through a large wound or becomes ill from bacterial infection, blood is lost and the body collapses. When this happens, crawling is impossible.

Thus far, however, the tent caterpillars that had survived the rigors of early spring were enjoying robust health. They were busy crawling, spinning trails of silk, constructing tents, and eating. Disease had not yet struck the colony, and few predators had attacked it. Bizarre as their mode of locomotion was, it carried them to all parts of the cherry tree. In weeks to come it would carry them on longer, more hazardous journeys. Eventually the entire marvelous machinery would be demolished by scavenger cells, as the caterpillar began to build within its body the muscles to power true legs that could really walk and wings that could free it from its earthbound existence.

IX

*'Tis true: there's magic
in the web of it.*
Shakespeare—*Othello*

Millions of years before Ariadne gave Theseus a sword and a clew of thread to guide his return from the depths of the labyrinth after he had slain the minotaur, tent caterpillars had anticipated the stratagem. For them, however, there was no enamored Ariadne. Each spun its own clew to lead it back home from the labyrinthine thickets of twigs in which it foraged.

Beyond the egg stretched a limitless unknown world, as limitless to the caterpillar, relative to its size, as the universe is to mankind. To venture into the unknown demands courage or ignorance, a goal whose position is known, or a guiding star to lead one back from positions known or discovered. Anyone can strike out on such a journey. Few can return. Successful explorers are they who have some clue as to the whereabouts of home and a compass to direct them there, or who, lacking knowledge of the location of home, follow a trail.

For those who would cruise the skies, there can be no lasting trail. Small insectan aeronauts, whose ships are themselves, can travel short distances close to the ground by following tangled skeins of odor blown and shredded by the gentlest of breezes. Thus, the longhorn beetle seeking a log in which to lay her eggs can orient to the resinous scents of newly felled firs and spruces. The long-tailed ichneumon wasp, that parasitizes the beetle grubs, follows the same tenuous trail.

Signs carry the same message for all who can read the language. Only the interpretation differs. To the beetle, the trail signifies

"tree"; to the wasp it denotes "grub." The air is laced with such trails, each written in a different chemical language. For the beetle in Gray's "Elegy," who "wheels his droning flight" where "drowsy tinkling lull the distant folds," the trail in all likelihood is the rich fecund odor of manure. For nocturnal moths, like the future incarnations of the tent caterpillars, the trail is an exquisite mélange of aphrodisiacs.

The larger feathered-flyers who sweep empty reaches of the sky are denied the comfort of trails. Scents are too ephemeral, too fragile, to withstand the ragtag sport of the wind. Birds rely upon their skill at reading the compass, whether it be the compass of the earth's magnetism, the compass of the sun, or the compass of the stars. In all probability, they employ all three, for the distances they travel are great, and many undertake the journey for the first time without experienced guides. How Spartan it is that species like some Pacific plovers first launch their young on the migratory journey from Alaska to Hawaii!

Creatures of the land need no compass for their ordinary comings and goings because there are substantial trails in abundance. The attributes of substance and permanence can even be an embarrassment. Trails may be followed by those for whom they are not intended. The hare may despair of the trail that it unintentionally leaves for the fox, and the squirrel despair of the spoor for the weasel. These are tracks deposited not by design, but by inadvertence. Trails left by design are usually more cryptic and private. Many animals blaze a trail with specific scents purposely deposited en route. Even the lowly ant marks its passage with the effluvia from very special scent glands. Tent caterpillars lay a more lasting trail.

On this first warm day of spring when hunger begot restlessness, the young caterpillars began to wander. Theirs was not the unbounded freedom of the bird in the sky or the fish in the sea, or even the antelope on the prairie. They were constrained by the twigs on which they crawled. Unlike the squirrel, whose field of exploration is limitless; so long as there are trees whose arboreal chasms can be bridged by flying leaps, these caterpillars can go only where branches lead. The crossroads are many, the turnings, beyond re-

membering. Unlike young gypsy moth larvae, that lower themselves into the air from long silken threads which act as balloons to disperse them far and wide, the tent caterpillars are treebound. Only when they have stripped a tree of all of its leaves do they forsake its security.

This particular day, four caterpillars, more venturesome than the rest, began to ascend the branch that arched upward from the egg mass. They crawled head to tail in a private little parade. Other small parades ventured onto side branches. Each leader swung its head rhythmically from side to side. From the region of the lower lip, a pointed spinneret extruded a thread of silk so fine that only the glint of the sun revealed its zigzag presence.

For all its fineness, the thread rivaled in complexity all other fibers, whether natural or synthetic. Constructed of a core of tough protein encased in a water-soluble gelatin, it combined strength, adhesiveness, and flexibility. Two long cylindrical glands extending almost the full length of the body were given to its manufacture. Assembly-line fashion, the protein was produced in the posterior part of the glands, the gelatinous casing in the middle, and the final molding in a press that fashioned the whole into a quick-drying thread. The outer coat of the thread caused it to stick tenaciously to the smooth bark. Even though the caterpillars had eaten nothing since birth but their eggshells and some spumaline, they were born with a generous supply of silk.

Without the microscopic silken ladder, the climbers would have had a difficult time. The bark of the smaller cherry branches shone with the smoothness of glass. Here and there small raised pores, lenticels, that permitted the stems to breathe, afforded some footing. Scars from the growth of other years also helped. The entire histories of the branches and stems were recorded in these scars. Whenever a leaf fell, it left behind a scar and, in it, smaller scars, marking the course of conducting tissues that had nourished the leaf during its brief summer of life. Wherever a bud had grown to extend the stem, it left behind the scars of the fallen scales which had protected it in its tender days.

Despite all of these footholds, the greater part of the stem presented a forbiddingly hard, smooth surface. The four caterpillars,

clinging to the silken strands with claws that slid and slipped on the naked bark, ascended like four Lilliputian mountain climbers roped together. Suddenly, for no apparent reason, the leader stopped as though confused, swung its head in wide sweeps, and then turned abruptly about, brushed rudely past its companions, and retreated rapidly down the trail.

No danger had threatened. No marauding ant or spider had appeared. An olive-backed thrush, scared by a foraging squirrel while poking in the forest litter, had taken temporary refuge in the cherry bush. The tremor of its landing had shaken the bush less than the puffs of spring breezes. No bird, even if caterpillars were its preference, would have bothered with the climbers at this period in their lives. Their minute size protected them. Too many would be required to fill an empty crop.

Elsewhere, the world lay serene. The many predators that would later make summer a time of carnage, a time when creatures would battle incessantly to protect and exploit their own patterns of living, had not yet come on the scene. True, the times were lean because of long winter fasts, long winter slumbers, or energy-sapping migratory flights, but a large portion of sustenance now was being provided by plants. Even though leaves were only beginning to green, buds were edible, last summer's nuts and seeds were still to be had, and the succulent new growth of forbs supplied acceptable salads. The strange behavior of the leading caterpillar did not portend danger.

Was it possible that the supply of silk had been exhausted? There are two simple methods for discovering how generous a supply of raw silk a young caterpillar has available at any one time. One is to shake the spinner from a twig so that it is forced to lower itself toward the ground. In response to repeated prodding, it pays out line until it is literally at the end of its rope. Another method, less coercive, involves placing it on a dark rubber ball and measuring the trail as it circumnavigates the ball again and again. Both experiments yield the same results, a thread from five to six feet long. The lead caterpillar had traveled about four feet.

The striking feature of its behavior was not so much that it stopped advancing when it had depleted its supply of silk, but that

it turned abruptly around and retraced its steps with a vigor suggestive of pressing business elsewhere. You might well argue that little choice of direction presented itself, that if one did not go up, the only other course was down. You might argue further that the downward trail was now well established, that the element of the unknown and its uncertainty did not exist. Certainly, downward did not offer as much opportunity for replenishing energy as did upward. Leaves, food, lay ahead, not behind. Would not the better part of wisdom have been to fall in behind a new leader and continue the ascent? And it is doubtful that gravity had suddenly acquired an overwhelming attraction, because if the branch happened to be horizontal, abandonment of the lead position did not constitute a downward course. Reversal, rather than downwardness, lay at the core of the behavior.

The new leader, who had been the second in line, after a moment of confusion during which it seemed to be assessing the situation, accepted the honor of laying trail. In time it too apparently exhausted the supply of silk and returned to the egg mass below, where like others before, it entered the nest or joined a small group of resting fellows. Another replaced it. As still other future leaders ventured up from below, they added their bit to the already existing trail at the rate of about eight centimeters per hour.

The trail became a two-way highway with surprisingly little confusion of traffic. The travelers seemed to recognize each other in passing, to the extent that they would sideswipe briefly, then continue on their separate ways. If they got shouldered off the trail, they marked the detour with a bit of silk, if any remained. Other kinds of disturbance that were caused by an alien insect or a prodding finger evoked more of an irritated response. Passing traffic, however, flowed smoothly, members of the ascending line swaying their heads rhythmically, but independently, side to side, those of the descending line hurrying by with a more preoccupied mien. Later, when the trail would have become more firmly established, all traffic, up as well as down, would move with an appearance of greater surety.

As with brooks and paths, the trail in time became even straighter. The meandering river in the nearby meadow had first followed a

course of least resistance among the hummocks, boulders, and weeds. At each bend it had carved away the outside bank, while the slower currents hugging the inside bank had dropped their burdens of silt. It appeared as though the brook were intent on making its course more tortuous, as the curves became sharper. Eventually, the curves had almost bent back upon themselves until some freshet caused the hastening waters to cut across narrow necks. Oxbows of quiet water had been left behind with each straightening of the course.

Similarly, with paths in woods and fields the small twistings and turnings become straightened as successive travelers, or a sole habitual traveler, cut corners, take short-cuts, straighten angles. Perhaps the meadow mouse in his grass-thatched path tires of the familiar route. Perhaps a path that is too well-used invites unwelcome attentions. Or perhaps, even the sheer novelty of exploration stimulates a change of course. Whatever the reasons, paths evolve as surely as brooks.

Just as the more direct portions of a path become most thoroughly trodden, so the more well-traveled portions of the caterpillars' trail accumulated most silk. They also accumulated a subtle scent. The same reinforcement occurs in a path in the woods. To human beings, creatures of vision and touch whose poverty-stricken noses are blind to the more subtle and evanescent scents of nature, a path is merely a physical thing. It is a convenience. To all the creatures of the forest who also traverse it, the path is an unending, ever-changing array of signposts. Scents and effluvia reveal who has passed by, how long ago, whether hurriedly or leisurely, and even whether in a courting mood or not. While the silken trail might not tell so detailed a story, it nonetheless told a story.

The gross vandalism of a finger smudged across the caterpillar trail disrupts traffic as surely as does removal of a bridge over a river. If the broken trail is replaced with silken threads removed from some other trail, traffic resumes as though nothing had happened. Nothing of the subtleness of the trail is revealed, however, by this mishap and its repair. If, on the other hand, the repair is effected with a length of spider silk or silk spun by gypsy moth caterpillars, such confusion ensues as to suggest that there is much more

to the trail than its mere physical presence. Indisputably, tent caterpillars are able to distinguish their own silk from that of others.

Beyond the fact that the trail has a chemical signature, little is known about it. With a highly volatile agent like pentane, the trail can be dry-cleaned of its distinctive odor. Within seconds of application, the solvent evaporates and leaves no residue. The materials it has extracted have been washed away completely. Pentane poured on a twig in advance of caterpillars laying down a pioneer trail, and completely evaporated by the time the vanguard arrives, does not deter the column one whit. Pentane poured on a trail in use obliterates it as far as caterpillars are concerned, without in any way altering the physical characteristics. The dry-cleaned silk carries no message. If silk is soaked in pentane, the odor is captured so the artificial trails can be painted on any surface. The evaporating solvent now leaves behind a recognizable residue.

The story does not end there. Among the caterpillars are those who will not be led, those free spirits who stray from the straight and narrow. When such as they approach a fork in the branch, they often leave the well-marked trail to strike out on their own trips of exploration. Boldness is tempered with caution, however, because they meticulously lay a new trail. Ties with home are not completely cut. Before the afternoon light on the first day had become wan, the cherry branch was a bifurcating network of silken trails, some much traveled, others mere byways. Those trails that ended in a profusion of buds were well-traveled and as luxuriantly cushioned in silk as would befit royalty. Trails ending at frost-killed twigs or unopened buds were poor, thin, and untraveled.

Trails leading to an abundance of food carried chemical information, revealing that at the end there was food to be had. The beauty of the sign was that it always remained up to date. The travelers, then, were the sign painters. They left a tell-tale scent on the trail as long as they returned well fed. When all the buds had been consumed so that late arrivals had to return from the twigs with empty stomachs, they did not reinforce the chemical. Gradually it began to fade so that the trail became less enticing and attracted fewer caterpillars. The lack of reinforcement accelerated fading. Before long, the trail was completely abandoned.

As spring slowly matured into summer, the unfolding leaves of the cherry could not grow rapidly enough to satisfy the expanding appetites of the caterpillars. A branch would be stripped, the trail would grow cold, new trails would be established. But all trails led home, and home was a remarkable tent.

X

This castle hath a pleasant seat;
The air nimbly and sweetly recommends itself
Unto our gentle senses.
Shakespeare—*Macbeth*

The final plowing and harrowing of spring had been completed several weeks past, and the cow corn showed four inches of green in long rows that undulated with the rolling valley. Apple blossom time had come and gone. There was time aplenty on the farm to tend those chores that had to be fitted into spare hours. The farmer had sprayed his orchard against codling moths and other early pests. Since petals had fallen, honeybees escaped poisoning because their attentions were now directed to wildflowers in the meadow. Beneficial parasites of pests were not so lucky; they perished with their hosts.

The farmer was doing one task of decimation by hand. From across the kelly green valley he could be seen moving slowly from tree to tree. He carried a pole in one hand and a bucket of kerosene in the other. From time to time he paused, dipped the pole in the kerosene, flamed something in the tree. He was burning new tent caterpillars' nests. Had he been asked, he would have commented that the tents not only signified potential damage to his trees, but were, in addition, ugly messes. One could not really disagree with his sentiments. Some situations are so intrinsically ugly that no effusion of romanticism or poetry can redeem them. What beauty can there be in a municipal dump, in a littered roadside, in disease, decay, and disorder? Pragmatism may explain some ugly things, but does not beautify them. Disease may be beautiful to the parasite that causes it, but hardly to the host that suffers it. That which does

not belong, though it may be beautiful in itself, is ugly when placed in the wrong context. That which violates order and the integrity or harmony of something else surely is ugly. In no way can the tents of the tent caterpillars be viewed as beautiful. They are exquitely constructed of finest gossamer, marvelously engineered, cunningly adapted to need, but monuments to ugliness. They are ugly first because they violate the harmony and form of the cherry tree as a shroud conceals a body. They are ugly as disease is ugly because they portend the disfigurement of the cherry. They are ugly in their final weeks as a municipal dump is ugly, as a house fallen to wrack and ruin is ugly.

While the tents were still small, their repulsiveness lay not so much in appearance as in the apparent negation of design and order. But until they had served their purpose, which was not the purpose of the cherry tree, they did have design, and function as well. Only after their purpose had been fulfilled did they personify, in existence rather than in potential, ugliness in the absolute. When their purpose had been fulfilled, they were bereft of form, devoid of function, the essence of decay. All of this potential was what the passerby saw in his or her mind's eye. No understanding of existing function could erase that vision. Nature is, after all, under no compulsion to conform to man's abstraction of beauty.

Obviously oblivious of these considerations, the young tent caterpillars, even before the buds had fully unfolded, began to build a tent in a sturdy crotch some distance below the egg mass. No thought directed the effort; no experience was called upon; none of them had ever built a tent before; no dreams of a completed edifice motivated the builders.

The choice of location made sense even though no sensible considerations entered into the decision. A base of operations should be centrally located. By placing the tent below the many branching twigs, the caterpillars had assured themselves equal access to many areas of the cherry. The paths from the tent branched in all directions. Had the nest been out at the tips of the branches, the caterpillars would have had to descend to a larger branch or trunk in order to explore other parts of the tree. On the other hand, the greater distances that would have had to be traveled each day would have

nullified any advantages of a central location. Indeed, the tent seemed to provide the best of all possible worlds.

An amazing feature of the construction was that there were no architects, no foremen, and apparently no communication among workers. It was as though a castle were being erected by a group of blind and dumb workers. Each individual caterpillar made its own contribution, undirected except by the constraints imposed by the contours of branches and whatever previous construction had been completed. Yet out of seeming chaos came order, because the tent gradually assumed form and unity from the multiple uncoordinated efforts. This structure was no tower of Babel.

Not everyone labored. The day began with a clear blue sky accentuated by a few nascent cumuli. By the time the sun picked out the colony of caterpillars, a dozen or so had begun to stir while the majority lay abed. Within an hour, however, the whole population became active. All but six wandered up some silken trail to satisfy their personal appetites for young cherry leaves. These six, more ambitious than the rest, set about to do a stint of tent weaving.

Their ambitions at the start were modest. The first tent would be no more than two or three centimeters in the largest dimension. The first caterpillar laid a trail about one centimeter long up one of three branches forming the crotch. At this point, it turned down and strung a thread that formed a short hypotenuse from the branch to the mat of silk upon which the colony had been resting that night. This was not accomplished in the manner of a spider who swings adventurously from the end of a thread until the pendulum carries it across the void. It was done by reaching as far across the angle as possible, as though laying a thread across a chasm that could be bridged by stretching. Meanwhile, others of the six were spinning similar small bridges from branch to silk, and from one of their co-workers' bridges to another. After a quarter of an hour, a rough framework extended from branch to branch. Now the six began spinning lines back and forth on the top of the framework until a solid silk sheet emerged. Thus, a ceiling was formed over the original silk floor upon which the colony was resting.

After working for an hour, the six terminated their labors. Several hours later they resumed and either finished the layer or began

a new one. In the course of the succeeding days the 'tween-deck scaffolding was demolished, leaving the floor and a ceiling just high enough above it to allow space for crawling.

Each day a small cadre, not always the same individuals, bore the brunt of the labor. It is remarkable that each succeeding group was able to build upon the work of its predecessors without creating a monstrosity totally lacking in unity and harmony. One is reminded in some meausre of ancient monuments such as Stonehenge, construction of which extended over centuries, with neither communication of plans nor of intent transmitted from one century to the next. The end product was nonetheless impressive in unity of design and execution.

As the tent grew and layer upon layer was added, other signs indicated that summer had arrived. The calendar said that officially it was three weeks distant, and, in deference to this astronomical timetable, the temperature fluctuated widely. One day the wind would be stilled; the sun would shine with an August intensity. The farmer at his work would pause often to wipe the perspiration from his eyes. The vegetation would add centimeters to its stature almost overnight. Then in a matter of hours, a cool wind would blow from the northwest, and all nature would shiver. From the chimney of the farmhouse, a ragged plume of smoke would reveal that an early morning fire had been lit to drive out the chill and dampness of night. Man, the thinking animal, had gained moments of mastery over his environment. He had constructed dwellings which not only sheltered him from the elements, but which could be maintained at an even temperature despite the vagaries of nature.

However powerful thought may be, other forces were at work that were able to achieve much the same ends for lesser creatures. The tents on the cherry were the products of thoughtless efficiency and adaptation. By now they were no longer simple shelters. Constant traffic and renovations within had turned each into a maze of compartments interconnected by small caterpillar-sized doorways. Each room became a pocket of stagnant air, the temperature of which depended upon location. Those on the sunny side of the

tent were warm; those on the shady side, cool. All were moist. In a sense, the tent offered crude air-conditioning.

On the morning that the chimneys in the valley breathed their tell-tale wisps of smoke, there was not a caterpillar to be seen on the cherries or on the nests. All were huddled in a compact mass in the warmest rooms. As the sun rose and turned back the shadowy covers of the tent, the warmest rooms became too warm for comfort, whereupon the caterpillars moved to cooler compartments. With increasing morning temperature, they changed progressively from warm rooms to cooler ones. At mid-morning, they departed for breakfast, but a chill legacy of winter stung the air so that they soon returned to the tent.

Leaves were slightly cooler than bark, and bark was cooler than the tent. Thus, a gradient of warmth led inevitably to the tent. On this clear dry morning, the caterpillars sought the warmth and moisture of the nest. Had the day been cloudy and humid, they would have congregated on the outside to bask in the sun.

By mid-morning, solar heating reached a maximum. After a lag of two hours, the air caught up so that it was at its hottest and driest peak. The caterpillars had earlier crawled out to the farthest twigs for lunch, but as these places soon became uncomfortably hot, the diners returned to the cooler rooms of the tent. Not long afterwards, however, the sun beating on the tent as on a greenhouse made it unbearably hot. The caterpillars boiled out and sought the shady side of the branch.

Within the limits of a normal summer day, the tent provided crude temperature control, but perhaps no more primitive than a drafty uninsulated frame house with only fireplaces to heat it. As long as the caterpillars had built-in reflex responses to heat and humidity, and silken trails as well to lead them home, they could enjoy a measure of comfort that their furry coats were too poorly designed to provide. Their close relatives, the forest tent caterpillars, who neither spin nor build tents, had to rely solely on their ability to follow temperature gradients from leaf to twig to trunk, and to huddle compactly when the temperature dropped.

There were other master builders who had perfected the art of air-

conditioning to a much higher degree. They were discovered this morning by a racoon who had strayed too far away from home the previous night, and had been compelled to seek a temporary day's lodging in a hollow of a sugar maple. The maple, a venerable survivor of the days when all of the forests in the valley had been felled to make room for pastures and crops, had been spared to serve as a post for a barbed wire fence. The fence had long since rusted, except for a short section in the trunk, and another section partially buried in the moss. Attempting to heal the wound caused by the wire, the maple had covered it with scar tissue and bark. An end of wire protruding from the moss passed close by a young cherry bush. A small tent in this bush housed an equally small colony of caterpillars. The moth that had chosen to lay her eggs there certainly had had no conception of supply and demand, because the caterpillars had already stripped the bush even though they were in but their second growth stage. Deserting skeletonized branches, they had encountered the barbed wire, spun a trail along it, and thus had come to a blueberry bush. Although it was not a preferred food, they journeyed there twice a day, always returning to the shelter of the nest.

In the maple tree, the racoon had found a hollow to its liking. The error of its choice did not become apparent until the sun warmed the trunk. Then, from a small crack leading from the hollow to another smaller and more secluded chamber, first two honeybees emerged, then three, then an angry mob. The racoon, all thought of sleep forgotten, scrambled from the hollow and fled.

The honeybees had kept themselves warm during the winter by sealing every crack, except the entrance, of their particular chamber with masticated plant gum. When winter cold had enveloped the maple, they had huddled. Those on the edge served as a living insulating blanket, while those in the center generated heat by restless movement and agitation. They had stoked their metabolic fires by eating the honey stores and by breathing more rapidly than usual. In this way, while the caterpillars were still insensible in their eggs, the honeybees had been able to maintain their hive at a comfortable twenty degrees centigrade, regardless of the cold outside.

With summer at hand, they now faced the same problem as did

the caterpillars, namely, how to escape heat prostration. In this respect they excelled as engineers. As the day grew warmer, some members of the colony stationed themselves at the doorway where, with beating wings, they fanned hot air out of the hive.

Even in early summer, however, unseasonably hot days frequently occur. On the day that the caterpillars' coolest rooms had at last become unbearably hot, the bees found that fanning alone could not keep ahead of the rising temperature. Some secret message was given to the foragers, whereupon they deserted the wildflowers for the barnyard. Here, diligent search rewarded them with a source of water. Earlier that morning, the farmer had been pumping water for the chickens. Beneath the rusty hand pump, a sizable puddle mirrored the clouds. Now and again, a drop falling from the spout shattered the image, but the miniature waves hardly disturbed the ring of bees drinking at the shallow shores. No sooner did one forager drink her fill, than another took her place. Returning to the hive, each spread a layer of water over the combs or hung droplets from the surface. Vigorous fanning enhanced evaporation so much that the internal temperature dropped several degrees. Meanwhile, the fanners at the entrance continued to promote circulation by exhausting a steady stream of hot air. How crude the caterpillars' tent by comparison!

Primitive though it was, however, the tent normally served the caterpillars well for all but the last days of their lives. If they completely stripped a tree of its leaves, they left the tent in search of another source of food. Unless approaching maturity, they spun trails across the ground. Finding a new tree, they either constructed a new tent, or returned between meals to the old one, or dispensed with a tent altogether. Being too lazy to build a new tent could cost them their lives, and frequently did, if unseasonably cold or rainy weather occurred.

Regrettably, tent caterpillars are scandalously negligent housekeepers. The tent that began as a glistening, white, airy castle, became a slum dwelling. Its walls were soiled with graffiti of excrement and discarded skins. Within some of the chambers, diseased corpses lay unattended. Unlike ants, who scrupulously remove the dead, the caterpillars continued to show a lack of com-

munity spirit. Occasionally, if a pellet obstructed the passage of a caterpillar through one of the galleries, it picked up the obstacle, but then merely cast it aside somewhere else.

With the passing days, wear and tear from within, hastened by the vandalism of the tenants, so weakens the fabric that rain finds its way in to hasten the rot. Tearing winds tugging at loose ends widen the rents and rips. Black-billed and yellow-billed cuckoos tear at the tent to feast upon the occupants. No repairs are undertaken by the caterpillars. After they have passed four-fifths of their lives and have entered upon caterpillar old age, they cease mending or enlarging the tent.

By late June, the tent has served its purpose. It had not been built as a monument to the future. The potential ugliness inherent in the early days of construction is realized. Not even function can redeem the appearance now. In abandoned orchards, along roadsides, and in hedgerows, wherever there are apples and cherries, soiled, tattered nests disfigure the scene. Even the second crop of leaves that the trees put forth cannot cover the ugliness. Some birds use snippets of soiled silk to help camouflage their nests, but for the most part, the wrecks are undisturbed. Many an autumn storm will blow before the trees will be cleansed.

XI

*My hold of the colonies is in the close affection which grows
from common names, from kindred blood, from similar privileges,
and equal protection.*
Edmund Burke—"Speech on the Economical Reform," 1780

To all appearances the fields and forests are boundless
from one hilled horizon to the other and beyond. Split-rail fences
and field-stone walls, built in straight lines so uncharacteristic of
nature, are but feeble human attempts to place limits on that which
has no real limits. They serve the unnatural but practical end of
keeping the cows from the corn, the sheep from the alfalfa, the
horses from the road. Their less tangible purpose is to mark bound-
aries signifying that the land is subdivided and hedged by the legal-
ities of ownership.

In nature there is no ownership. There may be territory, but that
is an entirely different matter. The red fox, no respecter of fence or
walls, ranges freely over meadow, pasture, orchard, and barnyard. It
owns not a single square foot of these nor recognizes the ownership
of others; nevertheless, it has a hunting territory, bounded only by
right of eminent domain, by tenuous marking scents, by its ability
to rout intruders. Its methods are shared by many carnivores and
herbivores alike. This was the basis for the philosophy of the American Indians who viewed lands and waters not as things to be
owned but as territory to be used, respected, defended.

In breeding time the birds too stake out their territories by proclamation and defend them by bluff. In the marshes the red-winged
blackbird proclaimed his sovereignty in song from the vantage of a
cattail and bluffed away trespassers by feints and threatening dis-
plays. Even the timorous chipping sparrow proclaimed his small
principality around a chosen white pine. Among the lesser crea-

tures the large field crickets sat belligerently at their doorsteps and sang away their competitors.

Territory, however fiercely defended against kin, was shared with strangers. The fox suffered no other fox in his domain, but the kestrel hunted the same preserve in the day, and the barn owl at night. The chipping sparrow shared the pine with a robin, a pewee, and a magnolia warbler. Across the territories of all wandered the vagabonds, the homeless, the restless. Killdeer ranged the furrows of plowed land, swallows skimmed the grass tips, woolly bears foraged where they wished, ants ranged far and wide. The field that the farmer thought he owned played host to myriad throngs of the unlanded gentry.

Across the free land the inhabitants, whether territorial or not, spaced themselves out, each according to his species. Even the plants that seemed so chaotically crowded were spaced according to their natures. Trees shouldered others to a comfortable distance by the shade of their crowns. Some plants impregnated the soil around themselves with exotic chemicals that discouraged the intrusion of close neighbors. Still other species staked out a claim to an area of soil by occupying it with root systems so binding that no others could invade.

Some organisms preferred the solitary life of the pioneer and others the society of their kin. Between the two extremes were to be found all degrees of communalism. The choice lay in the heritage of the species. In the world of nature the dedicated hermits still had to seek companionship, however ephemeral, at mating time; and to the solitary parasite there eventually came the time when it needed its host.

The need for companionship, whether a passing mood or the bond of a lifetime, could be irresistible. A flock of cedar waxwings had stopped over for a few weeks in a grove of hemlocks down by the brook. Among them was a single bohemian waxwing, strayed or blown by storms from its home in the west. It had instinctively sought out its nearest of kin in this eastern land of strangers. Among these it would find no mate. The taboos and genetics of species identity were too powerful. Its future was oblivion, but for the moment its need was satisfied and it belonged to the flock.

In a handsome roadside rum cherry, partially consumed by tent caterpillars occupying three enormous tents, a swarm of honeybees had temporarily staked out a territory which few dared to challenge. For them territory was that of the colony rather than of the individual. Each individual sacrificed her personal territory to the colony. With the bees individual space was reduced to the absolute minimum—that area occupied by the body.

Under the weight of those packed bodies, waiting for the scouts to return with news of a suitable place to hive, the branch bent like a fully drawn bow. The farmer, from whose hive the swarm had emigrated, intended to provide a new hive of his rather than their choosing. With a helper he was careful cutting away obstructing foliage in order to prune the laden branch. At the last moment, however, the partially severed branch broke under the weight of the bees. Hitting the ground the swarm exploded into a million angry workers among whom, somewhere, was the queen. The men, veiled and gloved, retreated hastily. If the queen was safe, the swarm would re-form and perhaps yield to a second attempt by the men bent on domesticating it. Without the queen it would cease to exist as an entity.

Just as some force causes the solitary species to eschew companionship, some bond holds societies together. The swarm of bees needs its queen. It needs not only her presence but the knowledge of her presence and the behavioral talent to react to it socially. The honeybees in the rum cherry represented a society of the highest order. The tent caterpillars in the same tree were a communal group of a lower order of organization, but at the very least they shared with the bees a penchant for living cheek to jowl. Were their destinies also obligatorily linked? Did they still possess qualities enabling them to survive as rugged individuals? What unifying force bound them together? They had no queen. But scatter them today, destroy their tent, and tomorrow they would once again be a compact colony.

It is revealing to observe what happens when individuals are separated. A few caterpillars brought into the house provided partial answers. They were established on cherry twigs standing in tumblers of water—one caterpillar on one, two on another, three

on the next, up to eight because only eight tumblers were available. Fresh twigs were supplied as needed.

During the week that followed the solitary caterpillar roamed and searched and nibbled but never conquered its restlessness. The individuals making up the pair also squandered much of their time wandering. Eventually they found each other and spun the merest suggestion of a tent where they huddled between meals. These meals were interrupted by much wandering. The trio did much better, constructing a minute tent and wandering less. The octet fared best.

At the end of the week members of the larger groups had attained the same size as their compatriots in the colony at large. The solitary caterpillar was a diminutive lost soul. Eventually it grew to full size, but growing was a painfully slow process. Had a caterpillar psychiatrist analyzed it, the diagnosis would no doubt have been insecurity, loss of identity, neurosis, or what not. The smaller groups were less neurotic and were intermediate in size.

Under normal circumstances restlessness is related to hunger and quiescence to satiety. Bodily contact reinforces the quiescence of satiety, and without it a full stomach alone seems unable to allow rest. Contact between meals must fulfill some other need also, because without it a solitary caterpillar feeds only fitfully. It is as though the caterpillar has something on its mind.

There is a bond in contact—a need to cuddle. It would be interesting to learn whether or not a surrogate caterpillar, a piece of fuzzy pipe cleaner for example, would serve. Is fuzziness alone sufficient and would other species of caterpillars serve, or are there species scents and recognition cues? Would a lonely tent caterpillar behave toward others as the bohemian waxwing did with the cedar waxwings? The experiments have not been done.

The most obvious binding force for the tent caterpillar colony is the silk. Although an empty tent will not bring rest to a solitary caterpillar nor a silken trail quench his wanderlust, trails do normally unite the world of the tent caterpillar. When all roads lead to Rome, it follows that they bind territory to a single locus, lead individuals to that locus however distantly they stray, and provide

communal cohesiveness. Destroy the lines of communication and you dissolve the unity of a people. Destroy the trails of the caterpillars, prevent their repair, and the colony ceases to exist as a unit. Communication and communal contact: these are the ties that bind.

Lacking a leader, lacking a queen, the tent caterpillars nonetheless are a colony rather than a crowd. It seems almost contradictory that a group of individuals acting independently can build a communal tent, lay out a community of trails, coexist cooperatively without direction or leadership, and maintain these relations for as long as is mutually beneficial. For only that long does the colony exist. Later it will disband and each individual go its own way.

Until the caterpillars leave home forever their lives are shaped by the demands of the colony. The tent, the trails, and the hypnotizing effect of bodily contact channel wandering to prescribed territory, select by exclusion those that respond in unacceptably set ways to crowding and temperature control, ostracize the unfit and the misfits, and even determine diet preference to some extent by binding to a single tree individuals that might otherwise wander away in search of greener pastures.

At the same time, as studies with western tent caterpillars have shown, the colony is determined by the individuals composing it. Each tent houses leaders and followers, active and sluggish caterpillars. The proportion of the two determines the ultimate success or failure of the colony. During the uncertain days of early spring a colony composed of too many sluggards dies, whereas a colony with twenty or more percent of active members survives. Even later, when the days are warm and foliage spreads in abundance, a colony may be so inactive that the caterpillars are too lazy to leave the tent to feed themselves.

Leaders among caterpillars are born, not made. They take no courses in leadership nor do they profit by experience. The first born are destined for the role. Every female moth enriches the first eggs that she lays with the best and most reserves. Later eggs get the tailings. The first eggs hatch first, the larvae are more robust and active, they venture first into the uncertainties and dangers of

the world, and they lay the first trails that the followers travel. In their next life they become the most active moths.

Of the many effects that the leaders exert on the wellbeing of the colony none is more momentous than their role in clustering. Any motionless individual can form the nucleus of a cluster; consequently, the more numerous the inactive caterpillars the greater the propensity for colonial clustering. Disbanding of a cluster, however, depends upon agitators; but the agitators must be on the fringe of the crowd, and they must be there in critical numbers. Active caterpillars have generally fed heartily. Situated on the fringe, they begin to dry out, as digestion is completed, sooner than caterpillars buried in the middle of the mass. Being of an active disposition they become restless. Restlessness is catching. A ripple of activity spreads from the edge to the center of the colony, and the group disbands, followers and leaders alike.

If the caterpillars at the edge of the colony happen to be inactive individuals, they hold the colony together longer. More time is spent lolling about, less time is spent feasting, the colony grows and develops more slowly. Without leaders to stir up the masses the whole colony lapses into apathy. Instead of consuming large quantities of the best foliage and moving on to bigger, better, and more strategically situated tents, they cling to the old neighborhood. Here, with debris accumulating and cadavers rotting in unused chambers, disease becomes rampant.

In contrast to the sluggish, malnourished, infected colonies, those with a high proportion of active caterpillars extend their trails farther, exploit the best vegetation, grow faster, fight off predators more aggressively, and generally enjoy better health. The risks are probably no greater, only different.

A high proportion of active caterpillars profits the colony. Membership in an active colony profits the individual. Furthermore, the benefits accruing to an active caterpillar are passed to future generations. The active caterpillar, well nourished, becomes the active moth. The active moth, like her sluggard sister, produces both good and bad eggs, but being herself better nourished she lays a great number of grade A eggs before running out of reserves.

The time-worn arguments as to whether society shapes the indi-

vidual or the individual, society, find no simple answer in the tent caterpillar colony. As in all societies the answer must be that the relationship is a reciprocal one. In this particular society, lacking learning and cultural inheritance and automated by genes, it is no wonder that the pattern set and found successful in its limited sphere millions of years ago has been frozen into genetic rote.

XII

That which the palmer-worm hath left, hath the locust eaten;
and that which the locust hath left, the canker-worm eaten;
and that which the canker-worm hath left
hath the caterpillar eaten.
Joel 1:4–11.

From horizon to horizon the countryside presented an appearance of utter serenity. The birds had tired of their matinal choruses and were attending to some quiet business in the trees, bushes, and ground litter. Several woodchucks were concealed in the meadow grass. On the cropped side of the fence separating the pasture from the meadow a mixed herd of Holsteins and Guernseys moved leisurely among the buttercups. The tent of the caterpillars was empty because they were all out at the ends of twigs finishing their afternoon meal. The sky was empty of birds except for the speck of a buzzard tracing circles against the clouds. The entire scene was one of peace, rest, contentment. The green of forest and field lay motionless. There was nothing to indicate that it lived, that billions of plant cells labored to transform the energy of the sun to form and structure, to extract water and salts from the soil, to pump fluids skyward against the pull of gravity. All of the activity of growth and metabolism proceeded in silence.

Plants are the silent ones. They know no travail at birth or anguish at death. Only animals sing at courtship, squeal in fear, whimper in pain, and snarl or growl in aggression. Perhaps it is the nature of predators, whether they prey on other animals or on plants, to be vocal. In eating, however, they are silent.

As the afternoon shadows lengthened the wind died, the aeolian harp was wrapped in silence. A slight movement at the edge of a bramble hedge resolved itself into the forms of a cottontail rabbit

surveying the meadow from the entrance of a well-trod tunnel. Its nose twinkled as it sniffed the still air. Soon it sat bolt upright, a picture of fear and nervousness. Its ears swiveled from back to front to back again as it listened for sounds of danger. From the distance came faintly but clearly the sound of a cowbell. A few man-made noises, wood on wood as a door slammed, or a tool banged against a wall, or the squeak of a wheel, signaled the usual end-of-day activities. A dog barked briefly. A hermit thrush in the deep woods essayed a tentative evensong. All of the familiar sounds of the close of day for diurnal creatures and the beginning of night for the dwellers of darkness carried clearly through the still air. They only served to accentuate the stillness.

The aura of quiet and serenity was illusion. What the cottontail did not hear, nor could anyone hear, were the millions of jaws chewing, boring, sucking, tearing the fabric of the green plants. Whether in forest or field, the silent labors of the plants were as silently being undone by the attacks of herbivores large and small, from minute mites and beetles to the magnificent white-tailed deer.

While the rabbit listened carefully before beginning its own attack on the plants, it also watched wide-eyed for tell-tale movements of stalking predators. Its eyes, situated back on the sides of its head, could scan forwards, sideways, backwards without the rabbit moving its head. No hawk sailed overhead; no fox hunted within view. All the meat-eaters had either finished their hunting for the day or not yet begun their night's predations. It was too early for the hunting cry of the great horned owl. The world appeared as safe as it ever was for the herbivores.

The cottontail hopped into the greenery of the open meadow. To the untutored eye the vegetation presented a uniform expanse of green. To the artist and poet it revealed shades and hues ever changing as the light yellowed in its fading. To the rabbit the meadow presented a different texture, a texture of scents and essences. Against a broad background of freshness hung a smell of lushness where the brook wove a diffuse path of humidity. There is no word to describe that particular smell. It can only be appreciated by a visual designation, just as there was a brown odor

of dryness at a higher point in the meadow where granite outcroppings caused the lush grass to yield to sere species. Against these patterns a finer-grained patchwork of odors clung close to the ground, to be appreciated only by the more finely tuned herbivore perceptions. As the rabbit nosed its way among the plants, it recognized the delicate odors emanating from the various species of leaves—again, odors for which the language finds no words. And even if the language were rich and flexible enough to rise to the occasion, how can one demand of a language that which cannot be perceived or even imagined? Had the rabbit a language, it would have coined words for the smells of dandelion, devil's paintbrush, and great plantain, of clover, wild mustard, jenny-creep-over-the-grass, fescue, and timothy, or sorrel, dock, daisy, mullein, and buttercup, of scores of other herbaceous plants that constitute a meadow.

With the millions of scent cells that line the olfactory nasal tissue of the rabbit it can discriminate as a connoisseur among these odors. As it sniffs the leaves, it eschews the clover and buttercup, nibbles the plantain, and devours the dandelion. Like all herbivores it nibbles and chews with almost frantic haste as though the plant would escape and no other be found to take its place. As with all herbivores its nose tells it what to bite, and its tongue tells it whether or not its nose has deceived it. So it is with the woodchuck that forages earlier in the day, the white-tailed deer that grazes in the evening and early morning, the stolid domestic cattle, the meadow mice and deermice, and even the porcupines in the forest.

So also is it with the lesser herbivores, the fuscous field crickets whose only vocations seem to be singing and eating, and the solid, shorthorned grasshoppers that have no time for the frivolity of music but chew their way methodically through grass and grain. Every green plant hosts its devouring hordes whether they be four-legged, furred, clawed, and hoofed, or six-legged and armored in cuticle.

Whatever the rabbit can accomplish with its millions of olfactory cells the herbivorous insect can achieve with a mere fraction of that number and a diminished brain to match. Of all of the plant-feeding insects, the aphids, beetles, sawflies, grasshoppers and

locusts, true bugs, and caterpillars, the caterpillars can probably smell their way successfully through the botanical world of scents with the greatest frugality of sensory equipment.

As diverse as caterpillars are in their habits they are the most conservative in their equipment. One sensory model has been adapted to all feeding needs whether the caterpillar, like the woolly bear, be a generalist feeder, to whom dozens upon dozens of plants are acceptable, or a specialist, like the caterpillar of the monarch butterfly that feeds exclusively on milkweed, or the caterpillar of the spicebush swallowtail that feeds only on spicebush and sassafras, or that of the black swallowtail that restricts its diet to plants of the carrot family, that of the pipevine swallowtail that eats only pipevine, or the middle-of-the-road caterpillar whose menu is selective but not rigidly restricted to one or two items.

In this category belong the tent caterpillars which, like the tiger swallowtail and many others, select their food from a choice variety of unrelated plants. Yet, not all chosen plants are equally relished. There are clear preferences. True, the female moth had been careful to seek out cherry, apple, and crabapple so that the caterpillars when they hatched had only to find the buds. Nonetheless, they had inherited an ability to distinguish one plant from another. Their skill in this respect and their innate likes and dislikes are revealed when you place a colony of newly hatched, unfed, naive caterpillars in a container with an assortment of leaves. As they encounter one leaf after another they ignore some, such as oak and maple, without a nibble. Birch and willow are sampled and eaten reluctantly if nothing else is available. Cherry and crabapple are attacked voraciously without a second's hesitation.

As young caterpillars living a highly organized colonial life they are selective feeders. Toward the end of their larval days their allegiance to colony wanes, they wander more frequently from the cherry tree, sometimes never returning. During this period they are nomadic grazers with less finicky appetites. Oak, which was unacceptable to them in their youth, is now consumed with equanimity. Maple remains unacceptable; they have not lost their discrimination completely.

Decades ago it was said by biologists that caterpillars possessed a

botanical sense, a sense for identifying plants as surely as does the professional botanist. What the nature of this botanical sense was remained a mystery until relatively recently, but even now the mystery has not been solved completely.

All caterpillars have well-developed senses of smell and taste; all, regardless of their way of life, have the same kind of sensory structures to serve them. Just below the eyes, near the corner of each mandible, is a small, stubby antenna. It is equipped with three pegs that house altogether sixteen olfactory sense cells. In addition there is a pair of fleshy lobes on either side of the lower lip. One lobe has eight minute olfactory pegs and the other, two pegs with a total of eight taste cells. Eight more taste cells inside the upper lip complete the inventory of chemical senses. It is with this small number of sense cells that the tent caterpillar is able to perceive the world of plant odors and taste as efficiently as the cottontail does with its much more liberally endowed sensory systems.

The tent caterpillar crawling up a twig eventually arrives at a green leaf. Many features of a leaf distinguish it from a twig: color, shape, texture, a thin layer of high humidity at the surface, temperature, green odor, and odors specific to the plant. Not all of these features are perceived to the same degree or are equally important in regulating behavior, but together they give a sensory picture of "leaf" and of a particular leaf. Temperature, texture, form, color, and humidity are probably the least important if considered separately. If a leaf is too fuzzy, too tough, dead, or wilted, the caterpillar refuses to sample it unless a long time has elapsed since the last meal. There are on the antennae some extraordinarily sensitive temperature receptors with which the caterpillar can distinguish a wilted leaf from a turgid one and an intact leaf from one from which a few fresh bites have been taken. The turgid leaf evaporates more water than the wilted one, and a freshly bitten edge evaporates more water than an intact one. In both cases, the drop in temperature accompanying evaporation can be detected.

A most important characteristic of green leaves is "green" odor, an essence of freshness. It is derived from subtle but universally distributed mixtures of alcohol-related compounds constructed of six atoms of carbon. This essence, probably more than any other

single characteristic, distinguishes for the caterpillar green food from non-food. By analogy with music this might be thought of as the continuum against which the melody is played. The melody in the plant is the species-specific odor that identifies this or that plant among all others. The basis of these characterizing odors is a unique blend of special compounds. Cherry leaves are chemically identifiable as a blend of benzaldehyde, described as the odor of bitter almonds, hydrogen cyanide, and other as yet unidentified compounds. Tent caterpillars are able to detect this complex essence and transmit the information to the brain. In some manner beyond our comprehension the brain is able to sort out the incoming sensory messages, to recognize the one that means cherry (or crabapple) and to issue commands to the mouthparts to begin chewing and swallowing.

Over the millions of years that tent caterpillars have been associated with plants, their sensory detecting apparatus and brain computing systems have come to be tuned to those plants that provide the best diet for survival and growth. Caterpillar ancestors that chose inadequate or toxic plants either died before becoming moths or became infertile. In either case they left no progeny to perpetuate their errors of judgment.

Such a beautifully tuned relationship has finally evolved between the tent caterpillars and their foodplants that both continue to survive. No better proof exists than that the countryside still abounds in tent caterpillars and wild cherries. There is also no better proof that caterpillars are protected from selecting lethal diets than the abundance of tent caterpillars and the immunity of many species of plants to their attacks.

Some plants are protected because they lack the magic essence to stimulate feeding, others because they are simply unpalatable, others because they have combined a particular scent or taste with toxic compounds. Specialist caterpillars are protected by the finickiness imposed by a highly discriminating sensory system and a brain programmed to be triggered by a message that the experience of millions of generations over millions of years has shown to be wise.

Generalist feeders possess the same discriminating senses but

have much more liberal brains, brains that will issue eating commands upon receipt of almost all messages. Such liberal attitudes can lead to mistakes. If the mistake is fatal, that is the end of the matter. If the mistake is such that the caterpillar only becomes temporarily acutely ill, and recovers, another talent comes into play. Woolly bears, for example, may consume some leaves that induce alternate convulsions and paralysis. Although the toxin itself may be tasteless and odorless, the convalescent caterpillar recognizes by some attendant odor that a particular plant upsets the digestion. Henceforth, at least for a while, that plant is avoided.

Like the caterpillar the cottontail is born with certain inherited likes and dislikes. Like some caterpillars it learns by unpleasant experience to avoid plants that have once made it ill. Unlike caterpillars, it acquires a taste for some plants through its mother's milk. The dairy cattle enjoy the same capabilities. Sometimes hunger occasioned by overgrazing or drought drives them to overcome their natural or learned distaste for less preferred plants. When the mechanisms for selection are thus pushed to unreasonable bounds, the penalty may be death.

By these devices different plants have their own coterie of herbivores so that nature's greenery is spared indiscriminate onslaught by the total herbivorous horde.

Offense and defense, ploys and gambits, a battle for survival between the plants and animals, but the battle was waged in silence. The unknowing eye saw only peace. It saw only beauty. But there was beauty too in the struggle, beauty in the stratagems and beauty in the exquisite balance of the contestants.

By nightfall the plants had ceased their synthetic labors. The tent caterpillars had retired to their tents. The woodchuck was asleep in its burrow, the cattle in the barns. Many of the crickets had begun a nighttime of singing that would occupy them into the wee hours of the morning so that little time remained for eating. Other herbivores were abroad in force. The cottontail had worked its way down to the farm's vegetable garden where it began to wreak havoc among the new lettuce plants. Over where the young bean plants were growing some cutworms that had passed the day curled up in the soil close to the stems awoke, stretched themselves, and began

the work of neatly cutting one stem after another. In a patch of violets along the edge of a lawn the day-sleeping larvae of the silver-spotted fritillary prepared themselves for a night's feasting on violets. A porcupine awoke from its slumbers in the top of a giant poplar and began browsing. Deeper in the woods the first of a colony of beavers revealed its presence as a widening silver V as it headed for a meal of bark in a grove of poplars.

During the night the plants would suffer. With the first light of day they would begin anew the task of repair and replacement. All but the annual herbs would outlive their attackers. The herbs would out-reproduce theirs.

XIII

A plague o' both your houses.
Shakespeare—*Romeo and Juliet*

A small girl about six years old was picking the season's first wild strawberries in the field behind her house. Squatting on her haunches, she popped one berry into her basket, one into her mouth, one into the basket. Increasingly often the basket was forgotten in favor of the mouth. In this pursuit she had been happily engaged for longer than she knew because time was of so little consequence on this warm June day. Her idyll came to an abrupt end, however, when she popped into her mouth a berry that had just been hastily abandoned by a stink bug.

While the small girl was trying to wipe away the tenacious odor that the contaminated berry had left on her fingers, her father was busy in the vegetable garden dealing with another bug. He was spraying the young blue hubbard squash vines that had been visited by close relatives of the stink bug, the squash bugs. At the same moment a group of tent caterpillars were engaged in a pitched battle with still other members of the tribe, carnivorous shield bugs.

Just as the herbivores were silently ravaging the green plants the carnivores were as silently exacting their toll of the ravagers. Civilization has so freed mankind of its predators that there remains little awareness of predation as an omnipresent feature of nature. In their civilized security people are inclined to forget that they too were once active and individual participants in this battle. Times have so changed that the occurrence of an attack on a human being by a dog, a grizzly bear, or a shark is now newsworthy. People are even inclined to think of plagues and pandemics as unnatural

events to which they have no right to be subjected. The black plagues of the eleventh and seventeenth centuries are, in many minds, historical oddities. The third greatest pandemic that the world ever experienced, the flu of 1918 and 1919, lives as experience in only a few fading memories. Yet, apart from the circumscribed arena of human experience, plague and predation are part of the warp and woof of nature. Any organism, plant or animal, may at some time find itself the victim, and there is no Boccaccio, Defoe, or Camus to chronicle its agony.

The tent caterpillars are no exception. What weather and weak genes had failed to accomplish in cropping the population, pathogens, parasites, and predators were already effecting. Before the predatory bugs arrived on the scene, when the caterpillars were still small, the wolves of the insect world, the marauding ants, had taken their toll. A large isolated cherry tree that had been left standing at the edge of the pasture as a boundary tree was one scene of the early carnage.

By virtue of its large crown the tree housed five tents of different sizes situated far enough from one another so that the respective occupants had not mingled, as they frequently do, to form a single large colony. In this instance independence proved to be the undoing of the two smallest colonies. They had been discovered by a colony of ants. An average hard-working ant required only six minutes to drag off a caterpillar and return for a second. Within one and one-half hours, the two tents, containing twenty-eight and thirty-seven larvae respectively, had been annihilated.

The only protection that the young caterpillars have from ants is to escape detection long enough to grow the long hair of middle age. The effectiveness of a shaggy coat is convincingly demonstrated by barbering. Regardless of size of body, once a caterpillar is shaved, it is attacked and dragged off squirming to the subterranean ant nest, there to be dismembered for a formicine banquet.

The large colony that had escaped notice of the ants had now been discovered by four shield bugs, and they were hungry. How they knew that food was to be had there is a mystery. Perhaps some subtle scent of caterpillars stirred primitive urges that caused the

bugs to be attracted to the site. In any case their instincts were true because they arrived just as the caterpillars were emerging for their mid-morning repast.

On a weight for weight basis a bug is no match for a vigorous caterpillar. If the two happened to meet head on, the bug retreated before the crawling juggernaut. Strength and mass, however, do not always prevail over stealth. The battle of predation often matches stealth with agility, and there is never any certainty about the outcome. One caterpillar, having bluffed a bug out of its path, for some inexplicable reason then paused. The pause gave the bug an opportunity to attack. It circled to the rear of the caterpillar, always facing it, extended its beak forward, approached cautiously, and with a rapier-like thrust pinked its intended victim. At the sting the caterpillar lashed sideways with its rear segments. The bug was jolted to one side but, recovering quickly, resumed the attack by attempting to pierce the caterpillar in the side. It was knocked off its feet by a hard blow from the caterpillar's head. For a moment it appeared disconcerted; however, it made still another attempt directed this time at the neck. The caterpillar's reaction came with such vigor that the bug was knocked head over heels off the tent. Victorious and excited the caterpillar headed up one of the silken trails to partake of its delayed breakfast.

The tent had become a miniature battleground. While the encounter just described was unfolding another skirmish was taking place in a different arena. Here on a broad expanse of silk a bug and a caterpillar approached on a collision course and butted heads. The caterpillar kept pressing so closely that the bug had no maneuvering room to bring its beak from the position of rest between its legs into an attack attitude. While the caterpillar continued to butt, two other large colony members crawled back and forth over the persistent bug webbing it as firmly into the tent as Poe's hero in "The Cask of Amontillado" had bricked his victim into the wall of the wine cellar.

Not all caterpillars were so lucky. Once a bug succeeded in inserting its needle-like stylets beneath the skin all resistance by the victim rapidly ceased, probably because a tranquilizing toxin had

been injected. Even without this refinement a caterpillar would be unable to dislodge its tormentor because the tips of the stylets could be curved outwards to form two hooks. These could be withdrawn only at the discretion of the bug.

One large caterpillar so hooked was drawn up along a stem by a bug only one eighth its size with no hold other than the stylets. Not only had the bug tugged the caterpillar loose from its footing, it had also torn loose from the silken mat two other bugs that had also penetrated the caterpillar and refused to release their hold. In short order all three sucked the caterpillar as dry as an empty wine skin, withdrew their beaks, and returned to the tent to impale another victim.

By mid-afternoon a small heap of dry carcasses littered the tent and eventually were blown to the ground, whence some ants dragged them off to extract any residual nutrients. At last the bugs departed and the surviving caterpillars, not missing their departed comrades in the least, continued their unperturbed existence.

From time to time attackers even more skilfull than the bugs slipped among members of the colony while they were packed in close siesta on the outside walls of the tent. Two species of graceful ichneumon wasps visited colony after colony, slipped a long ovipositor into an unsuspecting caterpillar, deposited an egg, and departed unmolested. For that caterpillar the future was fated. The parasite would slowly consume the body from within, always careful to spare the vital organs so that the caterpillar could continue to feed for two. Not until the time arrived for the caterpillar to spin its cocoon would death come. All of the organs that were destined to become the adult had been consumed. Of the caterpillar's own tissues nothing remained but a sheath of muscles enveloping nothing and a nervous system to operate them. The caterpillar's last act was to spin. And from that cocoon in the following spring would emerge, not a moth, but a graceful lethal wasp.

Other small wasps also found the caterpillars. What the wasps did not parasitize the flies did. Five different species of coarse, hairy tachinid flies shared the caterpillars among them, leaving eggs attached, which upon hatching would become grubs, gnawing at the tissues, sapping the energies, and consuming, as had the wasp

grubs, those organs that were to have been the wings, the legs, the muscles, and the ovaries of the moths.

The colony that had welcomed spring two hundred strong had been reduced to half its number. The survivors gave every appearance of robust health. The cherry tree, however, was to have its revenge. Throughout the coldest spells of winter, it had provided a resting place of a deadly virus. An impregnable triangle bound the virus, the caterpillar, and the cherry. When the young larvae took their first meals of cherry, they placed in their stomachs an agent as deadly as the black plague.

From the stomach the virus insinuated itself into the blood, thence into the cells of the skin, the tracheal air ducts, and the fat body, the caterpillar's equivalent of the liver. Proceeding into the cell nuclei the viruses multiplied. In the process they synthesized many-sided crystals, polyhedra, of unparalleled toxicity. Almost over night many colonies were devasted.

At first the infected caterpillars showed no symptoms. Each day they left the tent on schedule to feed. Each day more virus entered their bodies. A deadly tide was seeping through the tissues. With the passing days the viruses unobtrusively invaded more and more cells. Beginning about the fourth day the caterpillars, to all outward appearances the picture of health, lost their appetite. By the tenth day the magnitude of the epidemic was apparent. Ninety percent of the colony was dead or dying. The interior of each victim's body was completely liquified. The skin had become so dry and brittle that the slightest contact or movement of wind caused a rupture through which the morbid, brown, virus-laden humour oozed in a spreading contamination. Colony mates that had till then escaped infection were doomed.

Many colonies were visited by huge epidemics. There was no panic, no awareness, no quarantine, no attempt to escape. Life in the colony proceeded as mechanically as clockwork until the biological forces were halted. Tents were festooned with drooping skins drained of their contents. Here and there a caterpillar saved by some personal immunity survived. One of these survivors, having fallen from the tree while feeding, by sheer luck found another cherry nearby. Also by sheer luck this tree housed a colony that had

escaped the epidemic. To this colony came the survivor thoroughly contaminated with virus.

The deadly cycle began anew, but the members of the healthy colony were older and more robust than the new arrival. Although infected by the rapidly spreading virus, they survived. But those that managed to escape all other dangers long enough to become moths later in the summer left a lethal heritage. Immune though they were they carried in their cells the genesis of future epidemics. Among the cells infected were those of the developing adult ovaries. Here the viruses eventually entered the eggs. The sins of the parents were already visited upon the following generation. In the spring to come the new young caterpillars would not only consume the viruses that overwintered on the cherry, they carried within themselves their own death warrants.

As if that were not enough the unusually wet late spring had been kind to molds, mildews, smuts, blasts, and bacteria. A particularly virulent spore-forming species of bacterium struck many of the colonies. Borne by the fresh warm breezes filtering through the forests and hedgerows the spores invisibly dusted the cherries and all upon them. Unknowingly the caterpillars continued to feed. For the first twenty-four hours nothing appeared to happen. There were no symptoms. By the second day a subtle change in behavior was noticeable. Many caterpillars became extremely irritable. They regurgitated excessively. By the third day appetite was gone, diarrhea had set in, the body voided all fluid, became paralyzed and short and dry. The fourth day the caterpillars were dead, reduced to mummified repositories of bacterial spores. With devastating rapidity a robust colony had been wiped out completely. The labors of weeks had come to naught. The carefully constructed tent served no further purpose; the silken trails were deserted. Only the cherry tree continued to grow.

The enormous population of tent caterpillars had been cropped by weather, starvation, ants, bugs, parasites, fungi, viruses, bacteria, and misadventure in general. It had been a particularly trying year. Summer had hardly begun and the die had already been cast for the year to come. There would be fewer moths, fewer egg masses, and fewer colonies of the next generation. As with all

species, not excluding mankind, the reproductive potential, the birth rate, genetically built in, generated an inexorable tide. Only the competing tides of multitudinous species all striving for their share of the finite materials of the planet keep the flood of life at an even level. The flood was never placid, and the waves, sometimes ripples, sometimes huge swells, reflected the fierce underlying competitive struggles. The rise and fall of the tent caterpillar populations from one year to the next, from one phase of a cycle to another, was just one facet mirroring the perpetual competition.

XIV

Sometimes too hot the eye of heaven shines.
Shakespeare—Sonnet 18

There came a day when the fires of the insatiable appetites of the surviving caterpillars burned low and died. No longer did the odor of cherry stimulate. No longer did the silken trails beckon. Instead a great restlessness seized the caterpillars. Before many other creatures had even begun life or had savored the succulence of the summer's new growth, or had entered upon their period of productivity, the caterpillars were experiencing the twilight of their lives as caterpillars.

The world was barely into its annual adolescence. The first crop of hay was not yet ready for mowing. Fields still rioted with the yellow of buttercups and of dandelions, the red of devil's paintbrush, and the white of daisies. Few plants had set seed. Milkweed blossoms were just freshing the air with their fragrance. The occasional honeybees that tippled there were a mere token of the numbers that were to arrive later in the season. The monarch butterflies migrating northward from Mexico, generation by generation, were still hundreds of miles beyond the reach of the fragrance. Leaves spread their planes to the sun with no disfigurement to hide. Only a solitary, fireman-red, longhorned beetle nibbled here and there.

In this time of commencement the remaining tent caterpillars abandoned their nests, crawled down the trunks of cherry trees or, climbing upward to the tips of outermost leaves, arched their backs and sprang recklessly into space. A great wanderlust had seized them.

Peoples and animals usually begin to wander when drought, hunger, war, pestilence, or intolerable crowding intrudes upon

them. Seldom do they disperse in the luxury of unbounded space and bounty of food. The caterpillars had both. Yet, some inscrutable change had occurred. No longer where they hungry. The mystery of why they no longer had appetites for food was as profound as the mystery of the nature of their appetites in the first place, but no more or less fathomable than the puzzle of human appetites. In caterpillars the physiological machinery is partly understood if you conceive of these beasts as being wired like little machines, driven as surely as fate by the instructions that their genes have programmed into the nervous system. To a point their appetites for food can be explained quite satisfactorily by assuming that the odor and taste of cherry stimulate the sense organs, which in turn trigger the central nervous system to initiate biting. In this model, biting, chewing, and swallowing continue until internal receptors associated with the crop signal the brain that that organ is full. The signal for fullness counteracts the signals for smell and taste so feeding ceases. In this view, hunger becomes the absence of inhibition from the crop. The system, then, would be as simple and automatic as the thermostat regulating the furnace at home.

Perhaps it is as automatic as this, but there certainly are times when the system shuts down completely. When a caterpillar is ill from bacterial or viral infection or when it has encountered a sublethal dose of insecticide, or when it has the misfortune to eat some of a toxic plant, the system no longer operates. Even though the crop be empty and the aroma of food as strong as ever, the appetite is gone. As it is with ourselves at such times food has no allure. Similarly before each molt the caterpillar loses interest in food. Now, nearing the end of its life as a caterpillar, it eschews food even though all the links in the feeding system are intact. The external sense organs still send their messages documenting the presence of succulent cherry leaves, the internal sense organs still monitor the fullness of the crop. All is ignored.

Perhaps a clue is to be found in the observation that the periods before molting and the period that ushers in wandering share something in common. Both are periods in which the small hormonal tides that swirl around the organs of the caterpillar undergo a change. From the time of hatching one of three hormones that con-

trols the growth and development of the caterpillar ebbs and flows periodically. At the flood it stimulates the biochemical machinery leading to molting. Toward the end of life as a caterpillar, the juvenile hormone, which has kept the caterpillar young, ceases to flow. The molt stimulated by the molting hormone now results in the caterpillar becoming a pupa.

Perhaps these or other hormonal tides, washing the nervous system, stimulating neurons, or actually being secreted within the brain itself by cells that play the dual role of neurons and glands, perhaps these nullify the sensory messages about food, blunt the appetite, and urge the locomotor centers to continual activity. Certainly there is a correspondence between the first flood tide of the molting hormone and the beginning of wanderlust. The mystery deepens and becomes ever more complex. Whether the caterpillars are flogged on by irresistible hormones or by other forces, the result is that the once compact social colony is dispersed far and wide.

One caterpillar headed toward an ideal location for a cocoon, just where the bottom row of shingles overlapped a stone foundation. For it, life as a moth was not destined to be. As it crawled upward, various small obstacles impeded its ascent: a bit of mud, a fragment of dried grass stem stuck to the stone, a strand of spider silk. All of these were circumvented with little difficulty. Even the strand of silk posed no great physical obstacle, but it was sticky—and tenacious. The caterpillar thrashed as though annoyed. The thrashing instead of freeing it brought it into contact with another sticky strand. It also sent tremors humming up the thread, and the tremors were a signal.

Concealed under the overlap of shingles clung a spider. This spider was neither large nor formidable in appearance, but this spider was hungry. She had been lying in wait for three days tending her lines like a patient fisherman but had not even had a nibble. Hers was no elaborate orbed web glinting in the sun or bespangled with dew drops. It was more in the genre of some modern painters, representational of nothing and produced with no apparent design, although it did make some apologies to a crude platform and funnel. Crude as it was, however, it was functional. Its very random-

ness enhanced its utility. Threads were flung out haphazardly as though the spider had been doodling in three dimensions.

The spider clung in her dark recess gently holding a thread in one delicately sensitive leg. After three days, during which the thread vibrated only to wind or to aerial debris bumping against it, there came a twitch different from the rest. Just as the experienced fisherman knows whether his line is rubbing bottom, has been hit by floating weed, or is being nibbled by a fish, and whether a small fish or a large one, so the spider assessed the vibrations in her web. The first one alerted her but carried little information. The second one, more persistent and more vigorous, carried a different message. It said that some creature had blundered into one of the far-flung lines, and it indicated which line. At the third twitch the spider left her hiding, ran daintily out part way to investigate, and saw the caterpillar.

By now the caterpillar had blundered into several threads. Escape was still possible because the legs and abdominal prolegs with their rings of hooks had a firm grasp on the stone foundation. Furthermore, the powerful muscles running the length of the body exerted tremendous force considering their size. Where the caterpillar was powerful the spider was dexterous. Like a dog snapping at a bear she rushed in, spun another thread or two and rushed back to escape the thrashing. The caterpillar made the mistake of losing complete contact with the stone as it struggled to rid itself of the sticky threads. Deftly, with lightning speed, the spider spun a thread that nearly doubled the caterpillar head to tail. The caterpillar did the rest. As it twisted, the spider reeled in the slack. With no great effort on the part of the spider the caterpillar wound itself up on the shortening line till by its own exertions it brought itself under the overlap of shingles. There where it might have spun its cocoon it was wrapped in a cocoon not of its own making. It would never become a moth. Elsewhere along the house and in countless similar places of refuge, places where cocoons might shelter, spiders waited and took their toll.

Other caterpillars wandered farther afield. One emerged from a jungle of grass at the edge of a macadam road. How much easier the going would be here. Accelerating its crawling it hurried toward

the far side. One cannot help but wonder why caterpillars that encounter a road always cross and never set out down its length. Few animals succumb to the lure of an open road. Where human beings see a road as beckoning to a goal, most large animals probably see a road as a place devoid of concealment, a place where they are vulnerable. A road is probably avoided less because of its association with that predator man than because of its being a place of expo sure to unknown dangers. Only that most domesticated of all animals, the dog, trots along a road. A cat, an animal that is really not domesticated, that condescends to take advantage of man's hospitality, retains its independence and its primitive wild soul. It crosses roads.

Caterpillars cannot see the other side of a road. The length and breadth are probably equally blurred so the choice of crossing rather than following along would not seem to depend on vision or on any perception of exposure and danger. One theory has it that caterpillars respond to the radiant heat of the road in such a manner as to be equally warmed on both sides. Could they not be symmetrically warmed by walking lengthwise down the middle? Whatever the mechanism, they and numerous other small creatures cross by the most direct route.

The day was hot, the sky cloudless. The sun beat upon the black tar making it so hot that a barefoot boy heading for a trout brook chose to walk along the gravel shoulder. A woodchuck caught on the wrong side as the boy approached elected to scurry across to the safety of its burrow on the other side. Its passage was so fleeting that the pads of its feet were barely warmed by the time it reached the meadow toward which it was fleeing.

The heat had a different effect on the caterpillar. For one thing it hastened its progress. As with all insects, the higher its temperature the faster its locomotion. Nothing compares in speed and frenzy to the antics of ants on a hot pavement. The same ants that earlier in the cool of morning move sluggishly, sedately, deliberately, move at high noon with unseemly abandon. So too the caterpillar picked up speed. Like a fast freight train each segment from the rear seemed to push impatiently on the segment ahead of it till it

appeared that the caterpillar was propelled by its rear end at a speed over which it had no control.

Tragedy lay ahead. The tar was black and hot; the caterpillar was black. True, there were the white lines and blue spots as well as the long brown hairs; but the greater area of the body was black—or nearly black. It, like the tar road, began to get hot. With sunlight beating from above, the tar radiating heat from below, no cooling breeze, and no perspiration, the caterpillar began to experience heat prostration. When it was in its tent it could escape the heat by seeking the shade. The road was its desert, and only half of the transit had been completed. Now the temperature acted in reverse. Where it initially speeded up the crawling, it now began to slow it. Ever more slowly the caterpillar crawled. At last it stopped—and baked. It was not alone. Up and down the length of the road others, some tentmates, some strangers, came to similar ends. A sparrow from the hedgerow flew out to one of the corpses, pecked in a sampling way but found it not to its liking. Some ants also tried to drag away one of the corpses.

To a casual observer it would appear that every caterpillar went its own way to destruction or a future. The population was in fact crawling in all directions in the grass jungle and forest litter. This waywardness did not mean, however, that each was unoriented, that no guiding forces were at work. Despite appearances, intrinsic orderliness charted their crawling, but design could be discerned only on bare smooth surfaces. When the sun shone brightly in the cool of morning the caterpillars had not shaken off the previous night's chill. Body temperature remained below thirty-two degrees centigrade, and they moved toward the sun. As the mid-morning temperature rose so also rose body temperature. When it exceeded thirty-two degrees centigrade, the caterpillars altered their courses to a heading at right angles to the sun. By noontime body temperature had risen to slightly more than thirty-five degrees. Now they moved directly away from the sun. Orderliness, adaptively sensible, designed in such a way as to guard against heat prostration (except on black roads!), ruled the course on uniform surfaces.

Along the roadsides, in the fields, under the forest canopy, the ground is neither smooth nor bare. Different parts—twigs, leaves,

and soil, shadowed and sunlit—have different heat capacities. A caterpillar goes constantly from one temperature to another, and no two caterpillars are at the same temperature at the same time. Accordingly, since orientation is temperature influenced, the over-all effect on the wandering population is one of great confusion and randomness.

On some days the movement is truly random. These are the times when it becomes clear that the business of navigating involves two phenomena. Decision-making, which course to choose, is the first; the second is knowing whether one is accurately on course and how to remain on course. In some manner not understood the temperature of the body selects the course—toward the sun, away from the sun, at right angles to the sun. How then does the caterpillar mark where the sun is? The simple answer, that it sees the sun or at least the direction from which light comes, suffices as long as the sun is visible. What happens when the sun is obscured by clouds? Do the caterpillars stray off course? Do they go by dead reckoning? As is so often the case with apparently simple scientific questions, the answers here are not simple. If the sky is completely overcast, the caterpillars crawl in the direction of the brightest sector. If the sun is obscured, orientation and changes with temperature are less accurate than the ten-degree navigational precision of which the caterpillars are capable. With clouds passing overhead a curious thing happens. Billowy cumulus clouds marching sheep-like across the blue sky have little effect on the earthbound crawlers. Thin, high cirrus clouds, on the other hand, disrupt the orientation. In this odd difference is to be found the identity of the guiding clues. Cumulus clouds are huge swirling masses of water droplets. Cirrus clouds are cold masses of ice crystals. Each affects light differently. Ice crystals alter the plane of polarization, the plane in which the light waves vibrate, of light passing through; water droplets do not. Therein is the clue. Caterpillars, unlike ourselves, but like bees and many other insects, can see light and dark patterns in the blue sky. A rare capability that is not perfected in mankind is given to those eyes through which the visual world is but a fuzzy relief of smudged form. Nature's gifts are not distributed evenly, but each creature utilizes its talents to the fullest.

Survival depends upon this, and even then survival is not assured.

Away from the road, away from the crevices, the toll mounted. Some caterpillars met an unheroic end by being merely trod upon by larger creatures. Others fought epic but losing battles with packs of hunting ants and marauding wolfspiders. Still others proceeded apparently unmolested but carried within their bodies the seeds of their own destruction, infection and parasites acquired earlier in life. Of the several hundred caterpillars that had emerged from eggs a short four months earlier the survivors could be numbered in the tens. In one view nature had been prodigal and wasteful. Reproductive effort had been squandered recklessly. On the other hand, in a cosmic sense there had been no waste—much effort, but no waste. The reproductive efforts of the tent caterpillar moths clearly had evolved in such a way that unchecked they would populate the world with tent caterpillars, but there had been many failures. There is, however, little material waste in nature. The world derives almost unlimited energy from the sun, but its material riches must be guarded jealously because there is no replenishing them. Thus, although caterpillars that failed would never perpetuate their failures, their organic richness produced spiders, parasitic wasps, and even birds. Not even the caterpillars that had been crushed underfoot could be reckoned a total material loss.

Each in its own small inglorious way returned to the soil some of the elements that the cherry tree had removed. The contribution was minute indeed; nevertheless, it was multiplied to significance by the billions of similar small contributions by hosts of insignificant creatures. Material is recycled, reused, re-formed. At worst it may eventually become stored in unusable form, but it is not lost.

HERZLINGER, Regina. A managerial analysis of federal income redistribution mechanisms: the government as factory, insurance company, and bank, by Regina Herzlinger and Nancy M. Kane. Ballinger, 1979. 182p bibl index 79-14524. 22.50 ISBN 0-88410-368-4. C.I.P.

An examination of the administrative mechanisms used by different federal income redistribution programs: cash transfer, governmental provision of services, governmental loans or grants to the private sector to produce the services, and governmental insurance. The impact of each of the four mechanisms is evaluated and recommendations are made as to the best mechanisms in different circumstances. The text is followed by four sketchy case studies of the different administrative mechanisms. However, the criteria by which redistribution programs were selected for inclusion in the study seem questionable as is the classification of certain of the programs. No distinction is made between local government service provision programs financed by the federal government and programs directly operated by the federal government. Herzlinger and Kane then review evaluations of the impact of these programs. It is not clear whether they have reviewed all such evaluations or only a subset. The quality of the evaluative data is not discussed, and data are sometimes interpreted naively. There is an underlying bias in favor of the private sector (e.g., because certain mechanisms "are so widely used in the private sector, they

XV

How proud we are! how fond to shew
Our clothes, and call them rich and new!
When the poor sheep and the silkworm wore
That very clothing long before.
Isaac Watts—"Against Pride in Clothes"

Sooner or later to every creature there comes a time of supreme vulnerability. Most often that time is the beginning of life, in the softness of infancy and the defenselessness at birth. At the end of life too there is the final helplessness, but then it counts for little. For many animals the span between these two periods of crisis is beset only with the normal dangers associated with living in a fiercely competitive world. For others, however, mid-life may be posted with regular periods of exposure to exceptional danger, times when the usual defenses are down, when existence becomes most precarious.

When the hermit crab outgrows its borrowed snail shell and must transfer to a more spacious apartment, there is that moment during which its frail soft abdomen is exposed to scores of alert predators. When the mosquito struggles free from the pupal case of its aquatic youth at the mirrored surface of some placid pond, it clings weak and soft-bodied in full view of hawking dragonflies above and lean trout beneath. When the mallard drake loses his pinions at the close of the breeding season, he is denied the safety of the sky and must skulk around in the camouflage of a temporary "eclipse" plumage until new wing feathers enable him to fly once again. When caterpillars approach the time of transformation to moths and butterflies, they enter a period of immobility and softness, a time of vulnerability.

As this time draws near many caterpillars seek the shelter of

nooks and crannies where concealment and solitude offer the best hope of survival. Others burrow into the soil, or rotten logs, or the detritus of forest floor and field. Still others, especially those that will be butterflies, brazen it out in the open, relying solely on camouflage.

The monarch hangs its gold-spotted green chrysalis from the matching green leaves of milkweeds. It not only hazards exposure, it assumes added risk by its method of metamorphosis. Just the business of hanging upside down as it does would challenge the skill of any trapeze artist; but the monarch, having woven a pad of silk, fastens the hooks of its last pair of false legs into the tangle of silk, then daringly releases all other holds and swings head down. The trickiest moment arrives when the caterpillar sheds this skin that is attached to the leaf and must make a grab for a new hold before releasing the old. There is an element of magic in the process. According to some views the animal pinches a fold of old skin between segments of its abdomen while the tail end is swung out of and around the old skin. Hooks on the end of what is now the chrysalis reach for the pad, and the cast skin is released. Another view is that the lining of the old alimentary canal, which is also shed, is held tightly by the anal muscles until a new hold is secured. Only birth could be more hazardous than this transformation from caterpillar to chrysalis.

Other species are equally exposed but less daring. The black swallowtail anchors its chrysalis upright on a stem of Queen Anne's lace where a silken cummerbund around the middle secures it. The only riddle to be solved is similar to that of the man who wishes to remove his vest without removing his jacket. The swallowtail has to remove its caterpillar skin without coming out of its cummerbund. The clouded sulfur adopts a similar strategy and matches its color to the background, green against green, brown against brown.

The tent caterpillar seems to elect a more conservative approach. It spins a stout cocoon. In this it is not alone among weavers, although it can hardly be numbered among the aristocracy. One of the latter was discovered and domesticated for its royal role in the

year 2640 BC by the Lady Si-ling, wife of emperor Huang-ti. She cultivated the mulberry plants that these oriental spinners demanded as their food, she guarded the healthy from the diseased and deformed, she selected and set aside those cocoons from which the breeding stock would propagate. The others she ordered killed. And she invented a loom with which to spin again the silk that her servants patiently unraveled foot by foot, yard by yard, from the sacrificed cocoons.

Domestication produced an aristocracy of silkworms but exacted its toll. Domestication resembles in many ways the distillation of a fabulous elixir, a distillation by which a selected ingredient is brought to an immaculate state of purity. But instead of capturing the quintessence the process discards the impurities, the traces, the minor elements that give reality to the imaginary. After thousands of years of domestication the commercial silkworm has become the producer of exquisite silk, but the genetic distillation has left it a pallid beast, dependent upon those whom it serves, and with an appetite so finicky that few plants will sustain it.

Time and again enterprising innovators have attempted to find a hardier silkworm with the instincts of a gourmand rather than of a gourmet. All ventures have failed, none more spectacularly than the attempt to cross gypsy moth caterpillars with other silkworms. Descendants of the escaped gypsy moths that share forest and roadside foliage with tent caterpillars are the legacy of that flawed experiment. And both species, in common with the domesticated silkworm and nature's wild royalty, the giant silkworms with the mythological names, cecropia, luna, promethea, polyphemus, weave cocoons of the finest silk. Silk, the soft lustrous raiment of the high and privileged, begins as the tough armor of nature's lowly in their time of vulnerability.

The tent caterpillars weave rather middle-class cocoons; nevertheless, the process is as technically advanced for them as for the commercial and the giant silkworms. Those that had survived the wandering from the tent moved from one shelter to another as though seeking one uniquely suited to their satisfaction. Many a superior resting place was abandoned to a less suitable one. The

hormonally driven wanderlust kept the caterpillars constantly on the move. Some subtle interaction between stimuli of the world without and of the world within ordained that the wandering be restricted to areas offering shelter. In this respect shadow and body contact provided by narrow spaces probably played important roles. Thus, when the final signal to settle down came, chance found most caterpillars in a place of shelter, not necessarily the best, but better than none. Some stopped in the rough corrugations of oak trunks, some among the needles of white pine seedlings, others under rocks, and still others on fence posts, and even in old tents.

One caterpillar in particular ended its journey along the bead of a rain gutter. It stopped midway between the nest of a polistes wasp and that of a white-faced hornet. Neither insect paid the least attention to the intruder. The polistes, on the surface of her exposed paper comb—a miniature, flat, gray chandelier—was too busy checking her grubby young and guarding the comb against enemies real and imagined. The white-faced hornet zoomed in and out of her Japanese-lantern nest with scant attention to any of her neighbors. Even a large jumping spider that was exploring the rain gutter spared only a passing glance of her several staring eyes toward the caterpillar. Such unneighborliness was all the more welcome because inside the caterpillar a turmoil of preparation had usurped all activity including that of defense and flight.

Since its first glimpse of daylight upon hatching from the egg the caterpillar had spent all of its active hours alternately eating and spinning. It had ventured nowhere without spreading before it a secure footing; it had spun long exploratory trails to and from its tent; preceding each molt it had, if not in the tent, woven a mat to which it anchored its feet to gain leverage while crawling from its cast-off skin. Now it was about to undertake the most ambitious, the most sophisiticated tapestry of its life, a cocoon, a caterpillar's equivalent of an oriental rug.

Consider for a moment the weaving of an oriental rug. There is first the design, different for each household, each tribe, each village, each designer. An artist conceives the design. Then, instructions for executing it are committed to writing where they remain

unrealized until someone desires to have that rug woven. When the desire arises, material is gathered, a loom is found, and the instructions are sung to the weavers who, before the rug is completed, may tie as many as 31,680,000 knots according to the precise instructions. In its own humble way the loom too is important because it determines the length and width of the rug. The caterpillar ties no knots, but the successive stages of its endeavors resemble in principle those of the rug weaving.

The caterpillar's "desire" is a subtle change in its hormonal balance and the repleteness of its silk glands. So refined is the action of the body's internal environment that malfunction of a gland or incorrect timing can miscue the brain at most inopportune moments. Removal of the glands that release juvenile hormone can trick the brain into ordering a cocoon at the wrong time. Under these circumstances even a very young caterpillar will weave a cocoon, a very small cocoon. When all goes well, however, the glands do not shut down until the appointed time. At that time the flow of juvenile hormone ceases while the other hormones continue. At this level of understanding we know more of the caterpillar's "desire" than we do of the rug maker's. His ideas we cannot dissect and analyze.

The matter of pattern can also be analyzed. Unlike that of a rug the design of a cocoon is not a pattern of flowers, figures, medallions, or geometric shapes; nevertheless, the cocoon of each species of moth is characterized by distinct motifs. Furthermore, each cocoon is true to a particular species color. The complex double envelope of the cecropia is rich brown. The simpler cocoon of the tent caterpillar is white suffused with brilliant lemon yellow. These motifs are not ideas in the brains of caterpillars; they are the expressions of genes selected over the millennia, lost when false to the survival of the caterpillar, preserved when they serve well.

In the brain then, probably in that mysterious region called the mushroom bodies, are stored instructions for weaving a particular kind of cocoon. The instructions are stored as surely as if written. When the hormonal climate is propitious and the raw material for weaving gathered, the brain begins transmitting these instructions along the nerves to the muscles that will carry out the actual

weaving. As in the production of an oriental rug the whole process requires careful organization and teamwork, teamwork in this case of organs rather than of individuals. For the caterpillar on the gutter the loom was the crevice. All during its final larval stage its silk glands had been manufacturing silk and would continue to do so even as the cocoon was being woven. The brain began to issue commands. In response, with alternating stretch-bend movements of the forebody, the caterpillar began to weave. It stretched as far as it could to the edge of the beading, fastened a thread of silk, then bent backwards and sideways, all the while drawing thread from its spinneret. At the end of the bend it fastened the thread to the opposite side of the crevice. With hardly a pause it stretched again, pulling more of the same thread which was again attached to the bead. Moving back and forth like a shuttle it wove a loose maze of silk over and under its body. From time to time it shifted position so that its head came to lie where its tail had been. The thread remained unbroken.

Automatically, reflexly, the body swung rhythmically to and fro. If perchance some blundering bird or flying leaf rent the woven work, repairs were made with the same reflex movement as long as the weaving had not progressed to the next stage, the inner chamber.

So rigidly were the instructions programmed that there was no turning back once the weaving had entered a new stage. The shape of the crevice, the tactile stimulation that it offered at each contact, and the influence of gravity shaped the outside dimensions of the woven framework. The importance of the "loom" in this regard is dramatically revealed by placing a caterpillar in the impossible situation of a two-dimensional world, a flatland lacking ups and downs and corners and crevices. Such a world is provided by an inflated balloon. Placed inside of the balloon a caterpillar no longer performs stretch-bend exercises because there is nothing to which to attach the thread. Instead it performs figure-of-eight movements and succeeds only in weaving an endless flat mat covering the featureless surface.

Unlike the giant silkworms that weave cocoons with two envelopes, the outer loose and the inner compact, the tent cater-

pillar weaves but one. Its construction requires a different set of instructions. In giant silkworms instructions to begin the inner chamber are given when sixty to seventy percent of the available silk has passed through the spinneret. Somewhere somehow the flow of silk is metered, the brain is informed, a new set of instructions is issued. That caterpillar is now commanded to alternate its stretch-bend exercises with a swinging figure-of-eight pattern. Thus, by the addition of woof to the warp a smoothly lined chamber is woven.

As successive layers are added the silk glands gradually become depleted even though the synthesis of silk is continuing all the while. This is a providential circumstance because, the walls of the chamber being thickened from the inside, space is reduced and only by a reduction in the size of the caterpillar's body can cramping be prevented.

One critically important feature remains to be considered: a perfect, symmetrically compact cocoon would be the caterpillar's and the moth's death trap. The cecropia caterpillar provides a weakness at one end by altering the weave at that end. Here threads are woven in parallel, whereas the rest of the cocoon, including the opposite end, consists of a tighter cross weave. An emerging moth can more easily force its way through parallel threads than through any cross weave. This ingenious valve is generally woven at the upper end of the cocoon in response to a gravitational stimulus. The tent caterpillar, too, constructs an escape valve. In vertically oriented cocoons it is located at the upper end. In horizontally oriented cocoons one end is as good as the other.

At last the cocoon on the gutter was completed. For forty-eight hours without respite the caterpillar had been spinning one continuous thread and weaving it into its own personal oriental rug. As the end of the task approached the caterpillar smeared a yellow paste into the felt-work. This soaked into the woven silk much as the design of an oriental rug is accentuated by painting. There it quickly dried into a fine powder. The least disturbance of the cocoon caused little sulfurous clouds to swirl from the surface.

From within, the cocoon resembled the tight lattice of a Moorish seraglio. It permitted only a filtering of light and a fractured view of

the outside world. By now, however, the caterpillar no longer responded to the summery features of the world. Its juvenile life at an end, its task of construction completed, and its time of crisis imminent, it lay in its silken-armored chamber secure from all save strong-beaked birds that might seek it out and whatever parasites had been immured with their host.

XVI

Only two weeks remained till midsummer, and already millions of new faces looked out on the countryside. A pair of tufted titmice that had filled a birdhouse with a huge cushion of sphagnum moss cupped with a grass lining had produced there a quartet of young. Once these had acquired feathers there would be no doubt that they were titmice. In the nearby marsh, where each blueflag captured a piece of sky, the eggs of the peepers and toads had long since become tadpoles. Already sprouting legs, they were undoubtedly frogs and toads. From the warm soil of the fields un-counted legions of field crickets, cone-headed grasshoppers, short-horned grasshoppers, and bush katydids acknowledged their genealogy in the resemblance to their parents. In the pasture irrepressible spring lambs were clearly the offspring of the sedate ewes and pompous rams. Even in the farmhouse there had been a new arrival, a small thing but undoubtedly a human being.

What is so remarkable about the succession of generations? Is it not to be expected that young resemble their parents? How could it be otherwise? And yet resemblance is the exception rather than the rule. There are vast hordes of marine creatures and land-dwelling invertebrates producing eggs that give rise to young so bizarre that their lineage is apparent only to the expert. The newborn sea snail is nothing more than a stomach with a cap fringed with cilia that spin it through the water. Later it becomes a weird individual with enormous ear-like lobes. Young clams and oysters are similar.

Young starfish are transparent globs with numerous waving tentacles. The young sea urchin has no shell. Crabs are born with nightmarish forms.

In freshwater ponds young dragonflies have no wings, the diving beetles no hard-shelled wing covers. Young mayflies resemble caterpillars, craneflies look like grubs, mosquitoes are wriggling demons, flowerflies are maggots. On land young ants, bees, and wasps are helpless pasty grubs and young butterflies and moths only caterpillars.

It is given to some creatures to develop within the safety of the mother's body until the form and structure with which they will face the world have been fitted together and need only grow. It is given to other creatures to begin life in free-living eggs so liberally provisioned with yolk that they too face the world as miniature adults. But for each of these there are thousands that are thrust upon the world while they are still embryos, while they have not yet been molded into perfection. To these is posed the double task of succeeding in life as frogs, metaphorically speaking, while preparing to become princes.

Tent caterpillars are such schizophrenic beasts. At one and the same time they must be perfect caterpillars and potential moths. They must be prepared to meet all the challenges that life presents to a slow-moving, earthbound grazer while beginning the construction of the fragile furred body and gossamer wings that will offer them the freedom of the air. In one sense the caterpillar is but the womb of the moth; yet it must survive as an independent being until the day of metamorphosis.

Even on that day of hatching in the fickle climate of spring the caterpillar carries deep in the recesses of its body pockets of cells that are part of it and yet not of it. These are the cells that build the future wings, legs, eyes, antennae, and sexual organs of the moth. As the caterpillar grows they come to occupy ever more space in the body.

If a caterpillar were aware of anything, it would be aware of itself only as a caterpillar. Only in its genes is there a greater awareness, and this is at best a chemical predestination. The genetic heirlooms

bequeathed to the caterpillar include those genes that lead to its perfection as a caterpillar and those genes that lead to the perfection of the moth.

Chaos and monstrosity would surely result if there were no programming, if all genes competed to express themselves simultaneously. It has been provided, therefore, for genes to enter the stage of development on cue, for cells to carry out their missions at appropriate times, and to die when their tasks are completed. The making of an adult is as much a matter of programmed death as it is a matter of burgeoning life.

The caterpillar lying in its cocoon had one more pattern of behavior to execute as a caterpillar. This would be its last. As it had done four times before, it would molt. As before, the quiet period preceding the molt was a time for building new cuticle beneath the old; but this time a different internal environment prevailed. The juvenile hormone, that fountain of youth, had dried to a trickle. For the first time the molting hormone was free to direct developmental events by itself. Tissues now released from the thralldom of youth began to build according to new plans. The cuticle that was forming beneath the tired caterpillar skin no longer possessed larval characteristics. It was tailored to clothe adult tissues which had been growing while larval tissues were dying. It had contours to incase wings not yet formed, legs that would be long and slender, antennae that would be delicately feathery, and eyes that would see the world more clearly than did the primitive visual spots of earlier days. When the spasmodic labor of molting finally shucked the old skin and crumpled it at the bottom of the cocoon, the pupa would be revealed in this shiny new cuticle. To a startling degree it resembled a royal mummy case in shape as well as in the intimation of what kind of a creature it concealed. That intimation was, however, more of a promise of what was to come than a suggestion of what had been, as with mummy cases, or even of what was inside at the moment. What was inside at the moment was a nervous system, a few active muscles, and a pasty gruel of cells, chemical nutrients, and hormones. Three weeks would be required to reassemble the mess into a moth.

For three weeks the rest of the world moved ahead with its busi-

ness. A Baltimore oriole that had just hatched four eggs in its nest high in an oak, most of the preferred elms having fallen victim to Dutch elm disease, now began to search seriously for cocoons with which to feed the hatchlings. The cocoon on the gutter was safe since the oriole kept away from human activity. A wren, just commencing her nest in a box nearby, was neither ready for cocoons nor interested in them.

Farther afield other animals were at various stages of their own lives. A few butterflies had already left their caterpillar days far behind. A question mark butterfly burst out of the shadows of the woods into a patch of sunlight focused on a gray birch. As soon as it landed on one of the grizzled scars mottling the trunk it was as lost to view as though absorbed by the trunk. Under the seductive warmth of the sun, however, it unfolded its camouflaged wings and sprang into view as the tawny inner surfaces caught the light. Not far away a banded purple explored a dirt path. As it walked along it alternately folded and unfolded its wings. One instant the brilliant white bands flashed against the black purple ground-color of the wings; the next instant they disappeared. Higher in the forest of maples a tiger swallowtail cruised in grand swings among the hills and valleys of the canopy. Everywhere color and motion bejeweled the summer day.

As the serene days passed, warm and bright, the cells in the pupa rearranged themselves, divided to build adult tissue, reused the compounds released by dying larval tissue, supplied new muscle for the flaccid adult legs, filled the thorax with powerful flight muscle, pumped raw materials into the hormonal glands, added thousands of sense cells to the developing eyes and antennae, put together gonads to produce eggs or sperm, and fashioned an entirely new brain with new sets of instructions befitting an adult.

"When I was a child," wrote the apostle Paul, "I used to talk like a child, think like a child, reason like a child. When I became a man I put childish ways aside." As we all realize, becoming an adult is very much a matter of learning to behave like an adult. Nothing could be farther from the truth in the case of the moth. Adult behavior, under the influence of genes that have been awaiting the period of pupation, is wired into the brain as surely as a program in

a computer. Its execution awaits only the switch to activate responses to commands as they arrive.

As so often happens in the life of the insect, the activating switch is the build up of specific hormones. By the time that all adult organs have been formed the hormone that will activate metamorphosis, the eclosion hormone, has accumulated in the brain where it has primed the neural circuitry. At full time it causes the brain to send bursts of commands to the muscles of the abdomen to rotate and to twitch. These preparatory exercises continue for about thirty minutes. An equal period of rest then follows. A different set of commands is then issued. Like a drill master repeatedly shouting "one, two, three, four, one, two, three, four," the brain rhythmically commands the abdominal muscles to contract and relax in precise sequences. As a result a smooth wave of contraction moves from tail to thorax, at which point there occur vigorous shrugging movements of the wing bases. These continue until the thin brittle pupal skin ruptures. Immediately cessation of sensory stimulation by the tight enclosing skin shuts off the stream of commands. The moth awakens, struggles free of the skin and forces its way through the valve of the cocoon. It leaves behind its silken armor, its old skins, pupal and caterpillar, and whatever memories it might have had of life as a caterpillar. Nothing it experienced in that life follows it into its reincarnation.

XVII

Lucifer: All that hath life, hath equal part in life;
The hoary tree, the moth that lives a day—
Inre Madách—*The Tragedy of Man*

The moth encircled the cherry twig with a final packet of eggs. Gradually her grasp of the smooth bark relaxed, and she fell exhausted to the ground. The earth had completed one circumnavigation of the sun since she herself had begun life as an egg. In a sense the moth was back where she had started. Eggs had striven mightily to produce more eggs. Through the action of other organisms, parasites, predators, bacteria, viruses, all cycling to the same purpose, and through attrition by weather and accident, the net gain in eggs this year was zero.

To what end the exquisitely complex and meticulously synchronized machinery, the hormones, nerve impulses, biochemical syntheses, cell divisions, gene actions? Was the whole adventure a mere charade? An exercise in futility? Was it all a means of perfecting the cherry by challenging it to response? Or was the experience trial by ordeal to perfect the moths?

Perhaps only the genes counted after all. If the true essence of life on earth, the essence that perpetuated itself, was the genes, then the caterpillars were just one of many dwellings that genes had built for themselves. Their survival would depend on the success with which the body they constructed competed against the bodies of other genes for a place in the sun.

Still, to what end? How can one explain such an expenditure of energy to perpetuate motion for the sake of motion, perpetuation of moths not for the sake of moths but for the perpetuation? Is it unthinkable that so much planning should go for naught? Per-

haps there is no plan. Perhaps only chance rules events. On the other hand, could it be that there is an inexorable march of events, that events transpire as they do because it is impossible for them to do otherwise? In that case there is no chance.

In the cosmic view the caterpillar is more than an insect, more than a fragment of life. The tent caterpillar is the universe. One cannot ask the purpose of the tent caterpillar. One can only observe it and marvel. But in observing it one cannot escape wondering of the universe, why?

Glossary

andrenids: solitary nest-building bees that mine horizontally in banks, or vertically along roadsides and in fields.

backswimmers: bugs of the family Notonectidae that swim upside down beneath the surface of fresh water, backs shaped like the bottom of a boat; can inflict painful stings if handled.

bagworms: larvae of moths belonging to the order Psychidae, inhabiting portable cases covered with fragments of leaves, twigs, etc.

banded purple: a relative (*Limenitis arthemis*) of the viceroy butterfly; larvae feed on willow, birch, aspen, and poplar.

benzaldehyde: oil of almond found naturally in kernels of bitter almonds, peach pits, and related fruits.

biological clock: mysterious mechanism in cells and organs that marks off time, resulting in cyclic behavior, the most common being a rhythm of approximately twenty-four hours.

black knot: a virus disease of cherries and plums manifested as black, contorted swellings of stems.

blues: delicate blue butterflies of the family Lycaenidae, which also includes coppers and hairstreaks; its slug-shaped larvae feed on several species of buds and flowers.

brain hormone: a hormone secreted by a small group of secretory nerve cells in the brains of insects.

brain waves: rhythmic electrical activity in the brain, alpha waves occurring at the rate of ten per second in a quietly resting person with closed eyes; distinctive waves are found in epileptics and when brain tumors are present.

cabbage loopers: caterpillars of the subfamily Plusiinae injurious

to plants of the cabbage family; they lack two pairs of prolegs, and therefore they crawl with a looping gait.

cambium: in trees, a thin circumferential layer of dividing cells, which during each growing season adds a layer of woody material on its inner side toward the center of the stem or root and a layer of bark on the outer side.

caterpillar: the immature stage or larva of moths and butterflies.

central nervous system: in insects the brain and ventral longitudinal nerve cord plus its ganglia.

checkerspots: medium-sized black, brown, and yellow checkered butterflies; Harris' checkerspot and the Baltimore are two common eastern species.

chorion: the outer shell of the insect egg.

chrysalis: the pupa of butterflies.

cilia: short hair-like structures on the surface of cells that move the cell through the water or, if the cell is fixed, move water over it.

click beetles: beetles of the family Elateridae, able if placed on the back to leap into the air with a click and eventually to land right side up.

clouded sulfur: the common yellow butterfly of the meadows, *Colias philodice,* has wings with black borders; its larvae feed on clover.

cocoon: a covering composed partially or wholly of silk or other viscid fiber, spun or constructed by many larvae as protection for the pupa.

crochets: curved spines or hooks on the prolegs of caterpillars, often arranged in a crescent; also found on the posterior tip of pupae.

crop: the expanded portion of the alimentary canal which serves to receive and hold food.

cuticle: in insects the outer covering formed by secretions of the underlying epidermal cells, tanned and covered with a layer of wax.

cyst: a stage in the lives of some protozoa in which most water has been eliminated, the animal shrinks and rounds up, and a pro-

tective wall forms, enabling the protozoan to withstand adverse conditions.

diapause: a state of arrested development in insects occurring in egg, pupa, or adult under hormonal control which in turn may be influenced by day length, temperature, etc.; a means of surviving adverse conditions, especially cold.

ecdysone: a steroid hormone from the prothoracic glands of insects, first isolated by Karlson and Butenandt in Germany; from five hundred kilograms of silkworms twenty-five milligrams were obtained.

eclosion: escape of the adult insect from the cuticle of the pupa or of the last larval stage.

epidermis, hypodermis: the single layer of cells beneath the cuticle of arthropods (insects, crabs, ticks, etc.) responsible for secreting the non-living cuticle.

flagellum: a long hair-like structure at the surface of a cell, used in freely moving cells to move the cell through the water.

forbs: weedy plants such as dandelions, plantain, chickweed, etc.

fritillaries: brown butterflies checkered with black and spotted on the undersides of the wings with silver, of the subfamily Argynninae; their larvae feed on violets.

geometrid moths: a large family of moths deriving their name from the Greek word for land-measurer in reference to their larvae, measuring worms or inchworms.

glycerol, glycerin: syrupy, sweet, warm liquid used as sweetener, antifreeze, and in dynamite.

glycol: sweet, poisonous, syrupy liquid used in the manufacture of explosives and as an antifreeze.

horsehair worms: long, slender, hair-like round worms, sometimes found in tangled masses in fresh water; larvae parasitic in insects.

hydrogen cyanide: colorless, intensely poisonous gas with a characteristic odor.

ichneumon wasps: parasitic wasps of the family Ichneumonidae, some with ovipositors as long as four inches; most parasitic on Lepidoptera.

inchworms: larvae of moths of the family Geometridae, measuring worms, canker worms.

juvenile hormone: a hormone produced by a pair of glands (corpora allata) behind the brain of an insect, directing molts in the juvenile direction.

larva: the young insect which emerges from the egg in an early stage of development and differs radically from the adult.

leaf-miners: small flattened caterpillars that live between the upper and lower surfaces of leaves, leaving tell-tale broad or twisting galleries; many belong to the family Gracilariidae.

leaf-rollers: caterpillars living in leaves that they have rolled and tied with silk, belonging mostly to the families Tortricidae, Pyralidae, and Gelechiidae.

lenticels: small porous regions in a woody stem through which gases can move.

longhorn beetles: large beetles of the family Cerambycidae, some with antennae twice as long as the body; larvae bore in wood, sometimes emerging as adults several years later, after the wood has been made into furniture.

midges: small non-biting mosquito-like insects of the family Chironomidae; larvae one of the few insects containing hemoglobin in the blood; biting sandflies are in the same family but in the genus *Culicoides*.

molt, moult: to cast off the cuticle in the process of larval growth and pupal formation; also the time when this occurs.

mourning cloak butterfly: one of the earliest butterflies of spring, *Nymphalis antiopa*; wings dark brownish purple with yellowish border; its larvae feed on willow, elm, and poplar.

mushroom bodies: groups of nerve cells forming the largest and most highly developed integrating centers in the brains of insects.

ocelli: the simple eyes on the side of the head of larval insects.

ovipositor: the tube, sometimes longer than the body, by which female insects place their eggs in appropriate places for hatching.

pentane: a flammable liquid derived from petroleum, a solvent.

pistol-case bearers: small caterpillars of the family Coleophoridae

that construct cases of silk and debris, shaped like ancient pistols.

polarized light: light in which the direction of vibration of the transverse waves lies in a single plane as compared with natural sunlight in which the direction of vibration changes rapidly and randomly; light from blue sky is partly polarized.

polistes: a common slim-waisted wasp that nests under eaves and also under stones.

prolegs: fleshy, unjointed abdominal appendages of caterpillars and some sawfly larvae; false legs.

pupa: the resting inactive stage between the larva and adult of insects undergoing complete metamorphosis, e.g., moths, beetles, etc.

question mark butterfly: one of the early spring butterflies, *Polygonia interrogationis*; larvae feed on nettles, hopvine, elm, linden, and hackberry.

receptor: in insects a nerve cell that is stimulated by light, temperature, touch, or chemicals and transmits information to the central nervous system.

root hairs: tiny extensions of epidermal cells of roots which absorb fluids from the soil; a two-foot rye plant has fourteen billion root hairs.

scale insects: very small insects, also called bark-lice, many species of which are enclosed in specialized coverings and are immobile; shellac and the red dye cochineal are prepared from certain species.

Sepsidae: a small family of slender flies, principally scavengers, especially fond of fresh dung where they dance prettily, pirouetting with vibrating wings.

skippers: small, generally dull-colored, moth-like butterflies, so-called because of the rapid darting flight; common in fields.

spinneret: the modified opening of the salivary glands through which the silk glands of caterpillars open, in the form of a hollow spine.

springtails: small insects belonging to the ancient order Collembola, rarely exceeding five mm in length; common in soil, among herbage, under bark, etc.

spumaline: a frothy secretion of the accessory glands of tent caterpillar moths, spread over eggs where it solidifies.

stink bugs: also called shield bugs after the shape of the body; bugs of the family Pentatomidae.

supercooling: cooling a liquid below its freezing point without it becoming solid.

tabanids: horseflies and deerflies; males feed on nectar, females on nectar and blood; some common species with banded wings.

tachinid: short, stout, bristly flies, often found around flowers; eggs laid in caterpillars or on the foodplant where they are eaten by the caterpillar; some species deposit live larvae within the caterpillar.

thorax: the middle of three body regions (head, thorax, abdomen) of insects, bearing true legs and the wings when present.

tiger swallowtail: the common black-striped yellow swallowtail butterfly; larvae feed on cherry, birch, and many other plants.

tracheae: spirally reinforced internal elastic air tubes in insects.

Tyndall blue: a blue color obtained by the scattering of blue wave lengths of light by fine structures.

water bear: microscopic eight-legged animal with a brain and a pair of eyes, inhabiting standing water, moss, and other damp places; can live completely dried up for years.

water boatmen: underwater swimming bugs of the family Corixidae; swim with a quick darting motion by means of oar-like posterior legs; when not coming to the surface to breathe remain clinging to some object on the bottom.

whirligig beetles: elliptical steely-black beetles of the family Gyrinidae that swim in large congregations on the surface of water, spinning in graceful curves around and around one another.

white-faced hornet: a large black and white hornet, *Vespa maculata*, that builds large globe-shaped nests in trees and on buildings.

woolly bears: caterpillars of the family Arctiidae, the tiger moths, common species being the salt marsh caterpillar, the yellow woolly bear, the road-crossing brown and black *Isia isabella*.

References

GENERAL

Alcock, J. 1975. *Animal Behavior*. Sunderland, Mass.: Sinauer Associates.

Chapman, R. F. 1969. *The Insects*. New York: Amer. Elsevier Pub. Co.

Dethier, V. G., and Stellar, E. 1970. *Animal Behavior*. 3rd ed. Englewood Cliffs, N. J.: Prentice-Hall.

Roeder, K. D. 1963. *Nerve Cells and Insect Behavior*. Cambridge: Harvard Univ. Press.

Snodgrass, R. E. 1961. The caterpillar and the butterfly. *Smithsonian Miscl. Coll.* 143(6):1–51.

Tippo, O., and Stern, W. L. 1977. *Humanistic Botany*. New York: W. W. Norton.

Wigglesworth, V. B. 1939. *The Principles of Insect Physiology*. London: Methuen.

CHAPTER I

Blais, J. R. 1935. Effects of weather on the forest tent caterpillar *Malacosoma disstria* Hbn. in central Canada in the spring of 1953. *Canad. Ent.* 87:1–8.

Hanec, W. 1966. Cold hardiness in the forest tent caterpillar, *Malacosoma disstria* Hubner (Lasiocampidae, Lepidoptera). *J. Insect. Physiol.* 12:1443–49.

Le Baron, W. 1870. The apple-tree tent-caterpillar. *American Entomoligst* 2:143–46.

Snodgrass, R. E. 1922. The tent caterpillar. *Smithson. Rpt.* 1922:329–62.

Stehr, F. W., and Cook, E. F. 1968. A revision of the genus Malacosoma Hübner in North America (Lepidoptera: Lasiocampidae): Systematics, biology, immatures, and parasites. *U.S. Nat. Mus. Bull.* 276:1–321.

Sullivan, W. N., and Thompson, C. G. 1959. Survival of insect eggs after stratospheric flights on jet aircraft. *J. Econ. Ent.* 52:299–301.

Witter, J. A., and Kulman, H. M. 1972. Mortality factors affecting eggs of the forest tent caterpillar, *Malacosoma disstria* (Lepidoptera: Lasiocampidae). *Canad. Ent.* 104:705–10.

Williams, J. L. 1939. The mating and egg-laying of Malacosoma americana (Lepid: Lasiocampidae). *Ent. News* 50:45–50.

———. 1940. The anatomy of the internal genitalia and the mating behavior of some Lasiocampid moths. *J. Morph.* 67:411–35.

CHAPTER II

Hinton, H. E. 1946. Concealed phases in the metamorphosis of insects. *Nature* 157:552–53.

Mansingh, A. 1974. Studies on insect dormancy. II. Relationship of cold hardiness to diapause and quiescence in the eastern tent caterpillar, *Malacosoma americanum* (Fab.) (Lasiocampidae: Lepidoptera). *Canad. J. Zool.* 52:629–37.

Mansingh, A., and Smallman, B. N. 1967. The cholinergic system in insect diapause. *J. Insect. Physiol.* 13:447–67.

Palmer, J. 1976. *An Introduction to Biological Rhythms.* New York: Academic Press.

Schoonhoven, L. M. 1963. Spontaneous electrical activity in the brains of diapausing insects. *Science* 141:173–75.

Van der Kloot, W. G. 1955. The control of neurosecretion and diapause by physiological changes in the brain of the cecropia silkworm. *Biol. Bull.* 109:276–94.

CHAPTER III

Bucher, G. E. 1959. Winter rearing of tent caterpillars, *Malacosoma* spp. (Lepidoptera: Lasiocampidae). *Canad. Ent.* 91:411–16.

Hodson, A. C., and Weimman, C. J. 1945. Factors affecting recovery from diapause and hatching of eggs of the forest tent caterpillar, Malacosoma disstria Hbn. *Tech. Bull. Minn. Agri. Exp. Sta.* 170:1–31.

CHAPTER IV

Hodson, A. C. 1941. An ecological study of the forest tent caterpillar, Malacosoma disstria Hbn., in northern Minnesota. *Tech. Bull Minn. Agri. Exp. Sta.* 149:1–49.

Stacey, L, Roe, R., and Williams, K. 1975. Mortality of eggs and pharate larvae of the eastern tent caterpillar, *Malacosoma americana* (F.) (Lepidoptera: Lasiocampidae). *J. Kans. Ent. Soc.* 48:521–23.

Wetzel, B. W., Kulman, H. M., and Witter, J. A. 1973. Effects of cold temperature on hatchings of the forest tent caterpillar, *Malacosoma disstria* (Lepidoptera: Lasiocampidae). *Canad. Ent.* 105:1145–49.

CHAPTER V

Crozier, W. J., and Stier, T. J. B. 1928. Geotropic orientation in arthropods. I. Malacosoma larvae. *J. Gen. Physiol.* 11:803–21.

Dethier, V. G. 1942. The dioptric apparatus of lateral ocelli. I. The corneal lens. *J. Cell. Comp. Physiol.* 19:301–13.

———. 1943. The dioptric apparatus of lateral ocelli. II. Visual capacities of the ocellus. *J. Cel. Comp. Physiol.* 22:115–26.

Gunter, G. 1975. Obseravtional evidence that shortwave radiation gives orientation to various insects moving across hard surface roads. *Amer. Nat.* 109:104–7.

Sullivan, C. R., and Wellington, W. G. 1953. The light reactions of larvae of the tent caterpillars, *Malacosoma disstria* Hbn., *M. americanum* (Fab.), and *M. pluviale* (Dyar) (Lepidoptera: Lasiocampidae). *Canad. Ent.* 85:297–310.

Wellington, W. G., Sullivan, C. R., and Green, G. W. 1951. Polarized light and body temperature level as orientation factors in the light reactions of some Hymenopterous and Lepidopterous larvae. *Canad. J. Zool.* 29: 339–51.

CHAPTER VI

Truman, J. W. 1978. Hormonal control of invertebrate behavior. *Hormones and Behavior* 10:214–34.

Truman, J. W., and Riddiford, L. M. 1977 Invertebrate systems for the study of hormonal effects on behavior. *Vitamins and Hormones* 35:283–315.

Wigglesworth, V. B. 1970. *Insect Hormones.* San Francisco: Freeman.

CHAPTER VII

Byers, J. R. 1975. Tyndall blue and surface white of tent caterpillars, *Malacosoma* spp. *J. Insect. Physiol.* 21:401–15.

Simon, H. 1971. *The Splendour of Iridescence: Structural Colours in the Animal World*. New York: Dodd, Mead.

Hinton, H. E. 1973. Some recent work on the colours of insects and their likely significance. *Proc. Brit. Ent. Nat. Hist. Soc.* 6:43–54.

Fox, H. M., and Vevers, G. 1960. *The Nature of Animal Colours*. New York: Macmillan.

Mason, C. W. 1926. Structural colors in insects—I. *J. Phys. Chem.* 30: 385–95.

CHAPTER VIII

Gray, J. 1959. *How Animals Move*. Harmondsworth, Middlesex: Penguin Books.

Trueman, E. R. 1975. *The Locomotion of Soft-bodied Animals*. New York: Amer. Elsevier Pub. Co.

CHAPTER IX

Fitzgerald, T. D., and Gallagher, E. M. 1976. A chemical trail factor from the silk of the eastern tent caterpillar *Malacosoma americanum* (Lepidoptera: Lasiocampidae). *J. Chem. Ecol.* 21:178–93.

CHAPTER X

Davis, N. F. 1904. The apple-tree tent-caterpillar and its life history. *Commonwealth of Pennsylvania, Dept. Agri. Bul.* No. 120:1–44.

Sullivan, C. R., and Wellington, W. G. 1953. The light reactions of larvae of the tent caterpillars, *Malacosoma disstria* Hbn. *M. americanum* (Fab.) and *M. pluviale* (Dyar) (Lepidoptera: Lasiocampidae). *Canad. Ent.* 85: 297–310.

CHAPTER XI

Greenblatt, J. A., and Witter, J. A. 1976. Behavioral studies on *Malacosoma disstria* (Lepidoptera: Lasiocampidae). *Canad. Entom.* 108:1225–28.

Long, D. B. 1955. Observations on sub-social behavior in two species of Lepidopterous larvae, *Pieris brassicae* L., and *Plusia gamma* L. *Trans. Royal Ent. Soc. London* 106:421–37.

Wellington, W. G. 1957. Individual differences as a factor in population

dynamics: the development of a problem. *Can. J. Zool.* 35:293–323.

———. 1959 Individual differences in larvae and egg masses of the western tent caterpillar. *Can. Dep. For., Forest Ent. Path. Br., Bi-mon. Prog. Rpt.* 15(6):3–4.

———. 1960. Qualitative changes in natural populations during changes in abundance. *Can. J. Zool.* 38:289–314.

———. 1974. Tents and tactics of caterpillars. *Natural History* 83(1):64–72.

CHAPTER XII

Dethier, V. G. 1937. Gustation and olfaction in lepidopterous larvae. *Biol. Bull.* 72:7–23.

———. 1975. The Monarch revisited. *J. Kans. Ent. Soc.* 48:129–40.

Schoonhoven, L. M. 1972. Plant recognition by lepidopterous larvae. *Symp. Roy. Ent. Soc. London* 6:87–99.

CHAPTER XIII

Ayre, G. L., and Hitchon, D. E. 1968. The predation of tent caterpillars, *Malacosoma americana* (Lepidoptera: Lasiocampidae) by ants (Hymenoptera: Formicidae). *Canad. Ent.* 100:823–26.

Baker, A. D. 1926–27. Some remarks on the feeding process of the pentatomidae (Hemiptera-Heteroptera). *Report of the Quebec society for the protection of plants* 19:24–34.

Bucher, G. E. 1957. Disease of the larvae of tent caterpillars by a sporeforming bacterium. *Canad. J. Microbiol.* 3:695–709.

Clark, E. C. 1958. Ecology of the polyhedrosis of tent caterpillars. *Ecology* 39:132–39.

Glaser, R. W. 1927. Studies on the polyhedral diseases of insects due to filterable viruses. *Ann. Ent. Soc. Amer.* 20:319–41.

Green, G. W., and Sullivan, C. R. 1950. Ants attacking larvae of the forest tent caterpillar *Malacosoma disstria* Hbn. (Lepidoptera: Lasiocampidae). *Canad. Ent.* 82:194–95.

Smith, K. M. 1967. *Insect Virology.* New York: Academic Press.

Stacey, L., Roe, R., and Williams, K. 1975. Mortality of eggs and pharate larvae of the eastern tent caterpillar, *Malacosoma americana* (F) (Lepidoptera: Lasiocampidae). *J. Kans. Ent. Soc.* 48:521–23.

Stehr, F. W., and Cook, E. F. 1968. A revision of the genus Malacosoma Hübner in North America (Lepidoptera: Lasiocampidae). Systematics,

biology, immatures, and parasites. *U.S. Nat. Museum Bull.* 276:1–321.

Steinhaus, E. A. 1949. *Principles of Insect Pathology.* New York: McGraw-Hill.

Sullivan, C. R., and Green, G. W. 1950. Reactions of larvae of the eastern tent caterpillar, *Malacosoma americanum* (F.), and of the spotless fall webworm, *Hyphantria textor* Harr., to pentatomid predators. *Canad. Ent.* 82:52.

Warren, L. O., and Tadic, M. 1963. Parasites of the eastern tent caterpillar, *Malacosoma americanum* (Fab.) in Arkansas. *J. Kans. Ent. Soc.* 36:260–61.

Wellington, W. G. 1962. Population quality and the maintenance of nuclear polyhedrosis between outbreaks of Malacosoma pluviale (Dyar). *J. Insect Path.* 4:285–305.

CHAPTER XIV

Sullivan, C. R., and Wellington, W. G. 1953. The light reactions of larvae of the tent caterpillars, *Malacosoma disstria* Hbn., *M americanum* (Fab.), and *M. pluviale* (Dyar) (Lepidoptera: Lasiocampidae). *Canad. Ent.* 85:297–310.

Truman, J. W., and Riddiford, L. M. 1977. Invertebrate systems for the study of hormonal effects on behavior. *Vitamins and Hormones* 35:283–315.

Wellington, W. G., Sullivan, C. R., and Green, G. W. 1951. Polarized light and body temperature level as orientation factors in the light reactions of some Hymenopterous and Lepidopterous larvae. *Canad. J. Zool.* 29:339–51.

CHAPTER XV

Frankhauser, G., and Reik, L. E. 1935. Case building of the caddis-fly larvae *Neuronia postica* Walker. *Physiol. Zool.* 8:337–59.

Van der Kloot, W. G., and Williams, C. M. 1953. Cocoon construction by the cecropia silkworm. I. The role of the external environment. *Behaviour* 2:141–56.

———. 1953. Cocoon construction by the cecropia silkworm. II. The role of the internal environment. *Behaviour* 3:157–74.

———. 1954. Cocoon construction by the cecropia silkworm. III. Altera-

tion of the spinning behavior by chemical and surgical techniques. *Behaviour* 4:233–55.

CHAPTER XVI

Riley, C. V. 1879. Philosophy of the pupation of butterflies and particularly of the Nymphalidae. *Proc. Amer. Assoc. Adv. Sci.*, 28th meeting, 455–63.

Truman, J. W. 1971. Physiology of insect ecdysis. I. The eclosion behaviour of saturniid moths and its hormonal release. *J. Exp. Biol.* 54:805–14.

———. 1973. How moths "turn-on": a study of the action of hormones on the nervous system. *Am. Scient.* 61:700–6.

———. 1976. Development and hormonal release of adult behavior patterns in silkmoths. *J. Comp. Physiol.* 107:39–48.

———. 1978a. Hormonal control of invertebrate behavior. *Hormones and Behavior* 10:214–34.

———. 1978b. Hormone release of stereotyped motor programmes from the isolated nervous system of the cecropia silkmoth. *J. Exp. Biol* 74:151–73.

CHAPTER XVII

Dawkins, R. 1976. *The Selfish Gene.* New York: Oxford Univ. Press.

Appendix A
The first three centuries of studies of tent caterpillars listed in chronological order

1646 Reported in Baerg, W. J. 1935. Three shade tree insects. *U. Kan. Ag. Exp. Sta. Bull.* 317:20–27. In this article, Baerg also mentions observations made in 1649 and 1658.

1790 Dean, S. *New England Farmer or Georgical Dictionary*, p. 41.

1793 Fabricius, J. C. *Entomologia Systematica Emendata et Aucta.* Vol. 3, part 1, p. 433.

1796 Peck, W. D. Natural history of the canker worm. Boston: *Rules and Regulations of Mass. Soc. for Promoting Agriculture* 1(2):35–45.

1797 Smith, J. E., and Abbot, J. *The Natural History of the Rarer Lepidopterous Insects of Georgia.* 2 vols. For J. Edwards, London, printed by T. Bensley.

1841 Harris, T. W. *A report on the insects of Massachusetts, injurious to vegetation.* Cambridge: Folsom, Wells & Thurston. Pp. 265–71.

1852 Harris, T. W. *A treatise on some of the insects injurious to vegetation.* Pp. 373–75.

1856 Fitch, A. *1st and 2nd Report on the Noxious, Beneficial and other insects of the state of New York.* Pp. 181–98.

1865 Packard, A. S. Notice of an egg parasite upon the American tent caterpillar. *Pract. Ent.* 1:14–15.

1866 Walsh, B. O. Apple-tree caterpillar. *Pract. Ent.* 1:46–78.

1870 Le Baron, W. The apple-tree tent caterpillar. *Amer. Entomologist* 2:143–46.

1870 Riley, C. V. Third annual report on the noxious, beneficial and other insects of the State of Missouri. Sixth Ann. Rpt. State Bd. Ag. 1870, p. 120.

1878 Saunders, W. Observations on the eggs of *Clisiocampa sylvatica* and *americana. Canal Ent.* 10:21–23.

1890 Brumer, L. Insects injurious to young trees on tree claims. *Neb. Ag. Exp. Sta. Bull.* 14:106.

1896 Weed, C. The tent caterpillar. *N. H. Ag. Exp. Sta. Bull.* 38:47–59.

1898 Lugger, O. Tent caterpillars. *Minn. Ag. Exp. Sta. Bull.* 61:189–99.

Appendix B
Plants reported as food plants

(From H. M. Tietz, *Index to the Described Life Histories, Early Stages and Hosts of the Macrolepidoptera of the Continental United States and Canada.* 2 vols. Sarasota Fla.: Allyn Mus. Ent., 1972)

NOTE: Not all food plant records are reliable because in the wandering phase tent caterpillars are often found on plants that are unacceptable as food.

Acer sp.	maples
Acer saccharophorum Koch	sugar maple
Amelanchier canadensis (Linn.)	shadberry
Arbutus menziesii Pursh	arbutus tree
Berberis vulgaris Linn.	barberry
Betula sp.	birches
Betula alba Linn.	white birch
Betula nigra Linn.	river birch
Betula papyrifera Marsh.	paper birch
Betula populifolia Marsh.	gray birch
Carpinus caroliniana Walt.	American hornbeam
Carya sp.	hickories
Ceanothus sp.	redroot
Cercis canadensis Linn.	redbud
Cornus florida Linn.	flowering dogwood
Corylus sp.	hazels
Crataegus sp.	thorns
Fagus sp.	beeches
Fagus grandifolia Ehrh.	large leafed beech
Fragaria chiloensis (Linn.)	strawberry
Fraxinus sp.	ashes
Hamamelis virginiana Linn.	witch-hazel
Juglans nigra Linn.	black walnut

Liquidambar styraciflua Linn.	sweet gum
Malus pumila Mill.	apple
Nyssa sylvatica Marsh.	black gum
Ostrya virginiana (Mill.)	American hop-hornbeam
Photina arbutifolia Lindl.	christmas berry
Populus sp.	poplars
Populus fremonti S. Wats.	poplar
Prunus americana Marsh.	wild plum
Prunus armeniaca Linn.	apricot
Prunus ceraus Linn.	cherry
Prunus domestica Linn.	plum
Prunus domestica var. *galatensis* ex. Hook	prune
Prunus ilicifolia (Nutt. ex. H. & A.)	
Prunus pennsylvanica Linn.	pin cherry
Prunus persica (Linn.)	peach
Prunus serotina Ehrh.	rum cherry
Prunus virginiana Linn.	choke cherry
Pyrus communis Linn.	pear
Quercus sp.	oaks
Quercus alba Linn	white oak
Quercus stellata Wang.	post oak
Quercus velutina Lam.	black oak
Ribes sativum Reich.	currant
Robinia pseudo-acacia Linn.	black locust
Rosa sp.	roses
Salix sp.	willows
Sorbus americana (Marsh.)	American mountain ash
Tilia sp.	lindens
Ulmus sp.	elms
Vaccinium sp.	blueberries

Appendix C
Some parasites and predators of the eastern tent caterpillar

Tetrastichus malacosomae Girault (Eulophidae)
Oencyrtus clisiocampae (Ashmead) (Encyrtidae)
Telenomus clisiocampae Riley (Scelionidae)
Ablerus clisiocampae (Ashmead) (Eulophidae)
Anastatus sp. (Eupelmidae)

LARVAL AND PUPAL PARASITES

Hymenoptera
  Ichneumonidae
    Hyposoter fugitious (Say)
    Labrorychus analis (Say)
    Gellus tenellus (Say)
  Chalcididae
    Brachymeria ovata (Say)
  Braconidae
    Phobocampe clisiocampae Weed.
    Therion sp.
Diptera
  Tachinidae
    Euphororocera tachinomoides (Townsend)
    Exorista mella (Walker)
    Hyphantrophaga hyphantriae (Townsend)
    Lespesia archippivora (Williston)
    L. frenchii (Williston)
    Achaetoneura sp.

PREDATORS

Pentatomidae
  Podisus placidus (Uhler)
  P. modestus (Dallas)
Formicidae
  Various species